Exploring the Kuiper Belt and the Trans-Neptunian Region

Contents

Chapter 1

Kuiper belt

The **Kuiper belt** /ˈkaɪpər/ or /ˈkøypər/[1] (as in Dutch), sometimes called the **Edgeworth–Kuiper belt**, is a region of the Solar System beyond the planets, extending from the orbit of Neptune (at 30 AU) to approximately 50 AU from the Sun.[2] It is similar to the asteroid belt, but it is far larger—20 times as wide and 20 to 200 times as massive.[3][4] Like the asteroid belt, it consists mainly of small bodies, or remnants from the Solar System's formation. Although many asteroids are composed primarily of rock and metal, most Kuiper belt objects are composed largely of frozen volatiles (termed "ices"), such as methane, ammonia and water. The Kuiper belt is home to three officially recognized dwarf planets: Pluto, Haumea, and Makemake. Some of the Solar System's moons, such as Neptune's Triton and Saturn's Phoebe, are also thought to have originated in the region.[5][6]

The Kuiper belt was named after Dutch-American astronomer Gerard Kuiper, though he did not actually predict its existence. In 1992, 1992 QB_1 was discovered, the first Kuiper belt object (KBO) since Pluto.[7] Since its discovery, the number of known KBOs has increased to over a thousand, and more than 100,000 KBOs over 100 km (62 mi) in diameter are thought to exist.[8] The Kuiper belt was initially thought to be the main repository for periodic comets, those with orbits lasting less than 200 years. However, studies since the mid-1990s have shown that the belt is dynamically stable, and that comets' true place of origin is the scattered disc, a dynamically active zone created by the outward motion of Neptune 4.5 billion years ago;[9] scattered disc objects such as Eris have extremely eccentric orbits that take them as far as 100 AU from the Sun.[nb 1]

The Kuiper belt should not be confused with the hypothesized Oort cloud, which is a thousand times more distant and is not flat. The objects within the Kuiper belt, together with the members of the scattered disc and any potential Hills cloud or Oort cloud objects, are collectively referred to as trans-Neptunian objects (TNOs).[12]

Pluto is likely the largest and most-massive member of the Kuiper belt and the largest and the second-most-massive known TNO, surpassed only by Eris in the scattered disc.[nb 1] Originally considered a planet, Pluto's status as part of the Kuiper belt caused it to be reclassified as a dwarf planet in 2006. It is compositionally similar to many other objects of the Kuiper belt, and its orbital period is characteristic of a class of KBOs, known as "plutinos", that share the same 2:3 resonance with Neptune.

1.1 History

After the discovery of Pluto in 1930, many speculated that it might not be alone. The region now called the Kuiper belt was hypothesized in various forms for decades. It was only in 1992 that the first direct evidence for its existence was found. The number and variety of prior speculations on the nature of the Kuiper belt have led to continued uncertainty as to who deserves credit for first proposing it.

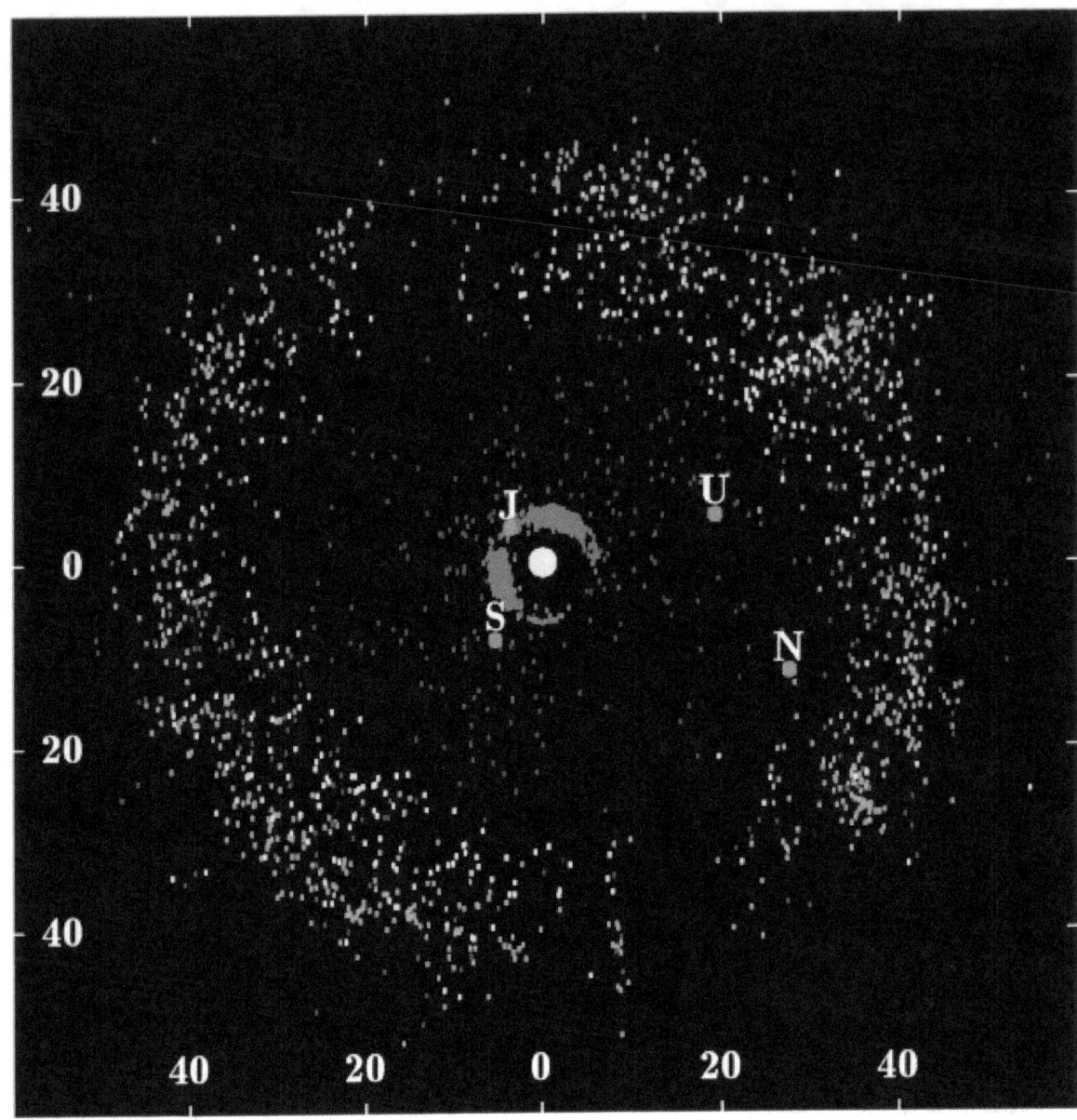

Known objects in the Kuiper belt beyond the orbit of Neptune. (Scale in AU; epoch as of January 2015.)
Distances but not sizes are to scale
Source: Minor Planet Center, www.cfeps.net and others

1.1.1 Hypotheses

The first astronomer to suggest the existence of a trans-Neptunian population was Frederick C. Leonard. Soon after
Pluto's discovery by Clyde Tombaugh in 1930, Leonard pondered whether it was "not likely that in Pluto there has come
to light the *first* of a *series* of ultra-Neptunian bodies, the remaining members of which still await discovery but which
are destined eventually to be detected".[13] That same year, astronomer Armin O. Leuschner suggested that Pluto "may
be one of many long-period planetary objects yet to be discovered."[14]

In 1943, in the *Journal of the British Astronomical Association*, Kenneth Edgeworth hypothesized that, in the region beyond
Neptune, the material within the primordial solar nebula was too widely spaced to condense into planets, and so rather
condensed into a myriad of smaller bodies. From this he concluded that "the outer region of the solar system, beyond

the orbits of the planets, is occupied by a very large number of comparatively small bodies"[15] and that, from time to time, one of their number "wanders from its own sphere and appears as an occasional visitor to the inner solar system",[16] becoming a comet.

In 1951, in an article for the journal *Astrophysics*, Gerard Kuiper speculated on a similar disc having formed early in the Solar System's evolution; however, he did not think that such a belt still existed today. Kuiper was operating on the assumption common in his time that Pluto was the size of Earth and had therefore scattered these bodies out toward the Oort cloud or out of the Solar System. Were Kuiper's hypothesis correct, there would not be a Kuiper belt today.[17]

The hypothesis took many other forms in the following decades. In 1962, physicist Al G.W. Cameron postulated the existence of "a tremendous mass of small material on the outskirts of the solar system".[18] In 1964, Fred Whipple, who popularised the famous "dirty snowball" hypothesis for cometary structure, thought that a "comet belt" might be massive enough to cause the purported discrepancies in the orbit of Uranus that had sparked the search for Planet X, or, at the very least, massive enough to affect the orbits of known comets.[19] Observation, however, ruled out this hypothesis.[18]

In 1977, Charles Kowal discovered 2060 Chiron, an icy planetoid with an orbit between Saturn and Uranus. He used a blink comparator, the same device that had allowed Clyde Tombaugh to discover Pluto nearly 50 years before.[20] In 1992, another object, 5145 Pholus, was discovered in a similar orbit.[21] Today, an entire population of comet-like bodies, called the centaurs, is known to exist in the region between Jupiter and Neptune. The centaurs' orbits are unstable and have dynamical lifetimes of a few million years.[22] From the time of Chiron's discovery in 1977, astronomers have speculated that the centaurs therefore must be frequently replenished by some outer reservoir.[23]

Further evidence for the existence of the Kuiper belt later emerged from the study of comets. That comets have finite lifespans has been known for some time. As they approach the Sun, its heat causes their volatile surfaces to sublimate into space, gradually dispersing them. In order for comets to continue to be visible over the age of the Solar System, they must be replenished frequently.[24] One such area of replenishment is the Oort cloud, a spherical swarm of comets extending beyond 50,000 AU from the Sun first hypothesised by Dutch astronomer Jan Oort in 1950.[25] The Oort cloud is thought to be the point of origin of long-period comets, which are those, like Hale–Bopp, with orbits lasting thousands of years.

There is, however, another comet population, known as short-period or periodic comets, consisting of those comets that, like Halley's Comet, have orbital periods of less than 200 years. By the 1970s, the rate at which short-period comets were being discovered was becoming increasingly inconsistent with their having emerged solely from the Oort cloud.[26] For an Oort cloud object to become a short-period comet, it would first have to be captured by the giant planets. In 1980, in the Monthly Notices of the Royal Astronomical Society, Uruguayan astronomer Julio Fernández stated that for every short-period comet to be sent into the inner Solar System from the Oort cloud, 600 would have to be ejected into interstellar space. He speculated that a comet belt from between 35 and 50 AU would be required to account for the observed number of comets.[27] Following up on Fernández's work, in 1988 the Canadian team of Martin Duncan, Tom Quinn and Scott Tremaine ran a number of computer simulations to determine if all observed comets could have arrived from the Oort cloud. They found that the Oort cloud could not account for all short-period comets, particularly as short-period comets are clustered near the plane of the Solar System, whereas Oort-cloud comets tend to arrive from any point in the sky. With a "belt", as Fernández described it, added to the formulations, the simulations matched observations.[28] Reportedly because the words "Kuiper" and "comet belt" appeared in the opening sentence of Fernández's paper, Tremaine named this hypothetical region the "Kuiper belt".[29]

1.1.2 Discovery

In 1987, astronomer David Jewitt, then at MIT, became increasingly puzzled by "the apparent emptiness of the outer Solar System".[7] He encouraged then-graduate student Jane Luu to aid him in his endeavour to locate another object beyond Pluto's orbit, because, as he told her, "If we don't, nobody will."[30] Using telescopes at the Kitt Peak National Observatory in Arizona and the Cerro Tololo Inter-American Observatory in Chile, Jewitt and Luu conducted their search in much the same way as Clyde Tombaugh and Charles Kowal had, with a blink comparator.[30] Initially, examination of each pair of plates took about eight hours,[31] but the process was sped up with the arrival of electronic charge-coupled devices or CCDs, which, though their field of view was narrower, were not only more efficient at collecting light (they retained 90% of the light that hit them, rather than the 10% achieved by photographs) but allowed the blinking process to be done virtually, on a computer screen. Today, CCDs form the basis for most astronomical detectors.[32] In 1988, Jewitt moved to the Institute of Astronomy at the University of Hawaii. Luu later joined him to work at the University of Hawaii's 2.24 m telescope at Mauna Kea.[33] Eventually, the field of view for CCDs had increased to 1024 by 1024 pixels, which allowed searches to be conducted far more rapidly.[34] Finally, after five years of searching, on August 30, 1992, Jewitt and Luu announced the "Discovery of the candidate Kuiper belt object" (15760) 1992 QB1.[7] Six months later, they discovered a second object in the region, (181708) 1993 FW.[35]

Studies conducted since the trans-Neptunian region was first charted have shown that the region now called the Kuiper belt is not the point of origin of short-period comets, but that they instead derive from a linked population called the scattered disc. The scattered disc was created when Neptune migrated outward into the proto-Kuiper belt, which at the time was much closer to the Sun, and left in its wake a population of dynamically stable objects that could never be affected by its orbit (the Kuiper belt proper), and a population whose perihelia are close enough that Neptune can still disturb them as it travels around the Sun (the scattered disc). Because the scattered disc is dynamically active and the Kuiper belt relatively dynamically stable, the scattered disc is now seen as the most likely point of origin for periodic comets.[9]

1.1.3   Name

Astronomers sometimes use the alternative name Edgeworth–Kuiper belt to credit Edgeworth, and KBOs are occasionally referred to as EKOs. However, Brian G. Marsden claims that neither deserves true credit: "Neither Edgeworth nor Kuiper wrote about anything remotely like what we are now seeing, but Fred Whipple did".[36] David Jewitt comments: "If anything ... Fernández most nearly deserves the credit for predicting the Kuiper Belt."[17]

KBOs are sometimes called **kuiperoids**, a name suggested by Clyde Tombaugh.[37] The term **trans-Neptunian object** (TNO) is recommended for objects in the belt by several scientific groups because the term is less controversial than all others—it is not an exact synonym though, as TNOs include all objects orbiting the Sun past the orbit of Neptune, not just those in the Kuiper belt.

1.2   Origins

The precise origins of the Kuiper belt and its complex structure are still unclear, and astronomers are awaiting the completion of several wide-field survey telescopes such as Pan-STARRS and the future LSST, which should reveal many currently unknown KBOs. These surveys will provide data that will help determine answers to these questions.[3]

The Kuiper belt is thought to consist of planetesimals, fragments from the original protoplanetary disc around the Sun that failed to fully coalesce into planets and instead formed into smaller bodies, the largest less than 3,000 kilometres (1,900 mi) in diameter.

Modern computer simulations show the Kuiper belt to have been strongly influenced by Jupiter and Neptune, and also suggest that neither Uranus nor Neptune could have formed in their present positions, as too little primordial matter existed at that range to produce objects of such high mass. Instead, these planets are estimated to have formed closer to Jupiter. Scattering of planetesimals early in the Solar System's history would have led to migration of the orbits of the giant planets: Saturn, Uranus and Neptune drifted outwards while Jupiter drifted inwards. Eventually, the orbits shifted to the point where Jupiter and Saturn reached an exact 2:1 resonance; Jupiter orbited the Sun twice for every one Saturn orbit. The gravitational repercussions of such a resonance ultimately disrupted the orbits of Uranus and Neptune, causing Neptune's orbit to become more eccentric and move outward into the primordial planetesimal disc, which sent the disc into temporary chaos.[38][39][40] As Neptune's orbit expanded, it excited and scattered many TNO planetesimals into higher and more eccentric orbits.[41] Many more were scattered inward, often to be scattered again and in some cases ejected by Jupiter. The process is thought to have reduced the primordial Kuiper belt population by 99% or more, and to have shifted the distribution of the surviving members outward.[40]

However, this currently most popular model, the "Nice model", still fails to account for some of the characteristics of the distribution and, quoting one of the scientific articles,[42] the problems "continue to challenge analytical techniques and the fastest numerical modeling hardware and software". The model predicts a higher average eccentricity in classical KBO orbits than is observed (0.10–0.13 versus 0.07).[40] The frequency of paired objects, many of which are far apart and loosely bound, also poses a problem for the model.[43]

1.3 Structure

At its fullest extent, including its outlying regions, the Kuiper belt stretches from roughly 30 to 55 AU. However, the main body of the belt is generally accepted to extend from the 2:3 resonance (see below) at 39.5 AU to the 1:2 resonance at roughly 48 AU.[44] The Kuiper belt is quite thick, with the main concentration extending as much as ten degrees outside the ecliptic plane and a more diffuse distribution of objects extending several times farther. Overall it more resembles a torus or doughnut than a belt.[45] Its mean position is inclined to the ecliptic by 1.86 degrees.[46]

The presence of Neptune has a profound effect on the Kuiper belt's structure due to orbital resonances. Over a timescale comparable to the age of the Solar System, Neptune's gravity destabilises the orbits of any objects that happen to lie in certain regions, and either sends them into the inner Solar System or out into the scattered disc or interstellar space. This causes the Kuiper belt to possess pronounced gaps in its current layout, similar to the Kirkwood gaps in the asteroid belt. In the region between 40 and 42 AU, for instance, no objects can retain a stable orbit over such times, and any observed in that region must have migrated there relatively recently.[47]

1.3.1 Classical belt

Main article: Classical Kuiper belt object

Between the 2:3 and 1:2 resonances with Neptune, at approximately 42–48 AU, the gravitational influence of Neptune is negligible, and objects can exist with their orbits essentially unaltered. This region is known as the classical Kuiper belt, and its members comprise roughly two thirds of KBOs observed to date.[48][49] Because the first modern KBO discovered, (15760) 1992 QB1, is considered the prototype of this group, classical KBOs are often referred to as cubewanos ("Q-B-1-os").[50][51] The guidelines established by the IAU demand that classical KBOs be given names of mythological beings associated with creation.[52]

The classical Kuiper belt appears to be a composite of two separate populations. The first, known as the "dynamically cold" population, has orbits much like the planets; nearly circular, with an orbital eccentricity of less than 0.1, and with relatively low inclinations up to about 10° (they lie close to the plane of the Solar System rather than at an angle). The second, the "dynamically hot" population, has orbits much more inclined to the ecliptic, by up to 30°. The two populations have been named this way not because of any major difference in temperature, but from analogy to particles in a gas, which increase their relative velocity as they become heated up.[53] The two populations not only possess different orbits, but different colors; the cold population is markedly redder than the hot. If this is a reflection of different compositions, it suggests they formed in different regions. The hot population is proposed to have formed near Jupiter, and to have been ejected out by movements among the gas giants. The cold population, on the other hand, has been proposed to have formed more or less in its current position, although it might also have been later swept outwards by Neptune during its migration,[3][54] particularly if Neptune's eccentricity was transiently increased.[40] Although the Nice model appears to be able to at least partially explain a compositional difference, it has also been suggested the color difference may reflect differences in surface evolution.[40]

1.3.2 Resonances

Main article: Resonant trans-Neptunian object

When an object's orbital period is an exact ratio of Neptune's (a situation called a mean-motion resonance), then it can become locked in a synchronised motion with Neptune and avoid being perturbed away if their relative alignments are appropriate. If, for instance, an object orbits the Sun twice for every three Neptune orbits, and if it reaches perihelion with Neptune a quarter of an orbit away from it, then whenever it returns to perihelion, Neptune will always be in about the same relative position as it began, because it will have completed 1 ½ orbits in the same time. This is known as the 2:3 (or 3:2) resonance, and it corresponds to a characteristic semi-major axis of about 39.4 AU. This 2:3 resonance is populated by about 200 known objects,[55] including Pluto together with its moons. In recognition of this, the members of this family are known as plutinos. Many plutinos, including Pluto, have orbits that cross that of Neptune, though their resonance means they can never collide. Plutinos have high orbital eccentricities, suggesting that they are not native to their current positions but were instead thrown haphazardly into their orbits by the migrating Neptune.[56] IAU guidelines dictate that all plutinos must, like Pluto, be named for underworld deities.[52] The 1:2 resonance (whose objects complete half an orbit for each of Neptune's) corresponds to semi-major axes of ~47.7AU, and is sparsely populated.[57] Its residents are sometimes referred to as twotinos. Other resonances also exist at 3:4, 3:5, 4:7 and 2:5.[58] Neptune possesses a number of trojan objects, which occupy its L_4 and L_5 points; gravitationally stable regions leading and trailing it in its orbit. Neptune trojans are often described as being in a 1:1 resonance with Neptune. Neptune trojans typically have very stable orbits.

Additionally, there is a relative absence of objects with semi-major axes below 39 AU that cannot apparently be explained by the present resonances. The currently accepted hypothesis for the cause of this is that as Neptune migrated outward, unstable orbital resonances moved gradually through this region, and thus any objects within it were swept up, or gravitationally ejected from it.[59]

1.3.3 "Kuiper cliff"

The 1:2 resonance appears to be an edge beyond which few objects are known. It is not clear whether it is actually the outer edge of the classical belt or just the beginning of a broad gap. Objects have been detected at the 2:5 resonance at roughly 55 AU, well outside the classical belt; however, predictions of a large number of bodies in classical orbits between these resonances have not been verified through observation.[56]

Based on estimations of the primordial mass required to form Uranus and Neptune, as well as bodies as large as Pluto (see below), earlier models of the Kuiper belt had suggested that the number of large objects would increase by a factor of two beyond 50 AU,[60] so this sudden drastic falloff, known as the "Kuiper cliff", was completely unexpected, and its cause, to date, is unknown. In 2003, Bernstein and Trilling et al. found evidence that the rapid decline in objects of 100 km or more in radius beyond 50 AU is real, and not due to observational bias. Possible explanations include that material at that distance was too scarce or too scattered to accrete into large objects, or that subsequent processes removed or destroyed those that did.[61] Patryk Lykawka of Kobe University has claimed that the gravitational attraction of an unseen large planetary object, perhaps the size of Earth or Mars, might be responsible.[62][63]

1.4 Composition

Studies of the Kuiper belt since its discovery have generally indicated that its members are primarily composed of ices: a mixture of light hydrocarbons (such as methane), ammonia, and water ice,[64] a composition they share with comets.[65] The low densities observed in those KBOs whose diameter is known, (less than 1 g cm^{-3}) is consistent with an icy makeup.[64] The temperature of the belt is only about 50 K,[66] so many compounds that would be gaseous closer to the Sun remain solid.

Due to their small size and extreme distance from Earth, the chemical makeup of KBOs is very difficult to determine. The principal method by which astronomers determine the composition of a celestial object is spectroscopy. When an object's light is broken into its component colors, an image akin to a rainbow is formed. This image is called a spectrum. Different substances absorb light at different wavelengths, and when the spectrum for a specific object is unravelled, dark lines (called absorption lines) appear where the substances within it have absorbed that particular wavelength of light. Every element or compound has its own unique spectroscopic signature, and by reading an object's full spectral "fingerprint", astronomers can determine what it is made of.

Initially, such detailed analysis of KBOs was impossible, and so astronomers were only able to determine the most basic facts about their makeup, primarily their color.[67] These first data showed a broad range of colors among KBOs, ranging from neutral grey to deep red.[68] This suggested that their surfaces were composed of a wide range of compounds, from dirty ices to hydrocarbons.[68] This diversity was startling, as astronomers had expected KBOs to be uniformly dark, having lost most of the volatile ices from their surfaces to the effects of cosmic rays.[69] Various solutions were suggested for this discrepancy, including resurfacing by impacts or outgassing.[67] However, Jewitt and Luu's spectral analysis of the known Kuiper belt objects in 2001 found that the variation in color was too extreme to be easily explained by random impacts.[70]

Although to date most KBOs still appear spectrally featureless due to their faintness, there have been a number of successes in determining their composition.[66] In 1996, Robert H. Brown et al. obtained spectroscopic data on the KBO 1993 SC, revealing its surface composition to be markedly similar to that of Pluto, as well as Neptune's moon Triton, possessing large amounts of methane ice.[71]

Water ice has been detected in several KBOs, including 1996 TO_{66},[72] 38628 Huya and 20000 Varuna.[73] In 2004, Mike Brown et al. determined the existence of crystalline water ice and ammonia hydrate on one of the largest known KBOs, 50000 Quaoar. Both of these substances would have been destroyed over the age of the Solar System, suggesting that Quaoar had been recently resurfaced, either by internal tectonic activity or by meteorite impacts.[66]

1.5 Mass and size distribution

Despite its vast extent, the collective mass of the Kuiper belt is relatively low. The total mass is estimated to range between 1/25th and 1/10th the mass of the Earth,[74] with some estimates placing it at one thirtieth of an Earth mass.[75] Conversely, models of the Solar System's formation predict a collective mass for the Kuiper belt of 30 Earth masses.[3] This missing >99% of the mass can hardly be dismissed, because it is required for the accretion of any KBOs larger than 100 km (62 mi) in diameter. If the Kuiper belt had always had its current low density these large objects simply could not have formed.[3] Moreover, the eccentricity and inclination of current orbits makes the encounters quite "violent" resulting in destruction rather than accretion. It appears that either the current residents of the Kuiper belt have been created closer to the Sun or some mechanism dispersed the original mass. Neptune's current influence is too weak to explain such a massive "vacuuming", though the Nice model proposes that it could have been the cause of mass removal in the past. Although the question remains open, the conjectures vary from a passing star scenario to grinding of smaller objects, via collisions, into dust small enough to be affected by solar radiation.[54]

Bright objects are rare compared with the dominant dim population, as expected from accretion models of origin, given that only some objects of a given size would have grown further. This relationship between $N(D)$ (the number of objects of diameter greater than D) and D, referred to as brightness slope, has been confirmed by observations. The slope is inversely proportional to some power of the diameter D:

$\frac{dN}{dD} \propto D^{-q}$ where the current measures[76] give q = 4 ±0.5.

This implies (assuming q is not 1) that

$N \propto D^{1-q} +$ constant a.

(The constant may be non-zero only if the power law doesn't apply at high values of D.)

Less formally, if q is 4, for example, there are 8 (=2^3) times more objects in the 100–200 km range than in the 200–400 km range, and for every object with a diameter between 1000 and 1010 km there should be around 1000 (=10^3) objects with diameter of 100 to 101 km.

If q is 1 or less, the law implies an infinite number and mass of large objects in the Kuiper belt. If $1<q\leq4$ there will be a finite number of objects greater than a given size, but the expected value of their combined mass would be infinite. If q is 4 or more, the law would imply an infinite mass of small objects. More accurate models find that the "slope" parameter q is in effect greater at large diameters and lesser at small diameters.[76] It seems that Pluto is somewhat unexpectedly large, having several percent of the total mass of the Kuiper belt. It is not expected that anything larger than Pluto exists in the Kuiper belt, and in fact most of the brightest (largest) objects at inclinations less than 5° have probably been found.[76]

Of course, only the absolute magnitude is actually known, the size is inferred assuming a given albedo (not a safe assumption for larger objects).

As of December 2009, the smallest Kuiper belt object detected is 980 m across. It is too dim (magnitude 35) to be seen by Hubble directly, but it was detected by Hubble's star tracking system when it occulted a star.[77]

1.6   Scattered objects

Main articles: Scattered disc and Centaur (minor planet)

The scattered disc is a sparsely populated region, overlapping with the Kuiper belt but extending to beyond 100 AU. Scattered disc objects (SDOs) have very elliptical orbits, often also very inclined to the ecliptic. Most models of Solar System formation show both KBOs and SDOs first forming in a primordial belt, with later gravitational interactions, particularly with Neptune, sending the objects outward, some into stable orbits (the KBOs) and some into unstable orbits, the scattered disc.[9] Due to its unstable nature, the scattered disc is suspected to be the point of origin of many of the Solar System's short-period comets. Their dynamic orbits occasionally force them into the inner Solar System, first becoming centaurs, and then short-period comets.[9]

According to the Minor Planet Center, which officially catalogues all trans-Neptunian objects, a KBO, strictly speaking, is any object that orbits exclusively within the defined Kuiper belt region regardless of origin or composition. Objects found outside the belt are classed as scattered objects.[78] However, in some scientific circles the term "Kuiper belt object" has become synonymous with any icy minor planet native to the outer Solar System assumed to have been part of that initial class, even if its orbit during the bulk of Solar System history has been beyond the Kuiper belt (e.g. in the scattered-disc region). They often describe scattered disc objects as "scattered Kuiper belt objects".[79] Eris, which is known to be more massive than Pluto, is often referred to as a KBO, but is technically an SDO.[78] A consensus among astronomers as to the precise definition of the Kuiper belt has yet to be reached, and this issue remains unresolved.

The centaurs, which are not normally considered part of the Kuiper belt, are also thought to be scattered objects, the only difference being that they were scattered inward, rather than outward. The Minor Planet Center groups the centaurs and the SDOs together as scattered objects.[78]

1.6.1   Triton

Main article: Triton (moon)
 During its period of migration, Neptune is thought to have captured a large KBO, Triton, which is the only large moon in the Solar System with a retrograde orbit (it orbits opposite to Neptune's rotation). This suggests that, unlike the large moons of Jupiter, Saturn, and Uranus, which are thought to have coalesced from rotating discs of material around their young parent planets, Triton was a fully formed body that was captured from surrounding space. Gravitational capture of an object is not easy; it requires some mechanism to slow down the object enough to be caught by the larger object's gravity. Triton may have encountered Neptune as part of a binary (many KBOs are members of binaries; see below); ejection of the other member of the binary by Neptune could then explain Triton's capture.[80] Triton is only 14% larger than Pluto, and spectral analysis of both worlds shows that their surfaces are largely composed of similar materials, such as methane and carbon monoxide. All this points to the conclusion that Triton was once a KBO that was captured by Neptune during its outward migration.[81]

1.7   Largest KBOs

See also: List of the brightest KBOs

Artistic comparison of Pluto, Eris, Makemake, Haumea, Sedna, 2007 OR$_{10}$, Quaoar, Orcus, and Earth.
(
This box:

- view

- talk

- edit

)

Since 2000, a number of KBOs with diameters of between 500 and 1,500 km (932 mi), more than half that of Pluto (diameter 2370 km), have been discovered. 50000 Quaoar, a classical KBO discovered in 2002, is over 1,200 km across. Makemake and Haumea, both announced on July 29, 2005, are larger still. Other objects, such as 28978 Ixion (discovered in 2001) and 20000 Varuna (discovered in 2000) measure roughly 500 km (311 mi) across.[3]

1.7.1 Pluto

Main article: Pluto

The discovery of these large KBOs in similar orbits to Pluto led many to conclude that, bar its relative size, Pluto was not particularly different from other members of the Kuiper belt. Not only did these objects approach Pluto in size, but many also possessed satellites, and were of similar composition (methane and carbon monoxide have been found both on Pluto and on the largest KBOs).[3] Thus, just as Ceres was considered a planet before the discovery of its fellow asteroids, some began to suggest that Pluto might also be reclassified.

The issue was brought to a head by the discovery of Eris, an object in the scattered disc far beyond the Kuiper belt, that is now known to be 27% more massive than Pluto.[82] (Eris was originally thought to be larger than Pluto by volume, but the *New Horizons* mission found this not to be the case.) In response, the International Astronomical Union (IAU), was forced to define what a planet is for the first time, and in so doing included in their definition that a planet must have "cleared the neighbourhood around its orbit".[83] As Pluto shared its orbit with so many KBOs, it was deemed not to have cleared its orbit, and was thus reclassified from a planet to a member of the Kuiper belt.

Although Pluto is currently the largest known KBO, there is at least one known larger object currently outside the Kuiper belt that probably originated in it: Neptune's moon Triton (which, as explained above, is probably a captured KBO).

As of 2008, only five objects in the Solar System (Ceres, Eris, and the KBOs Pluto, Makemake and Haumea) are listed as dwarf planets by the IAU. However, 90482 Orcus, 28978 Ixion and many other Kuiper-belt objects are large enough to be in hydrostatic equilibrium; most of them will probably qualify when more is known about them.[84][85][86]

1.7.2 Satellites

Of the four largest TNOs, three (Eris, Pluto, and Haumea) have satellites, and two have more than one. A higher percentage of the larger KBOs possess satellites than the smaller objects in the Kuiper belt, suggesting that a different formation mechanism was responsible.[87] There are also a high number of binaries (two objects close enough in mass to be orbiting "each other") in the Kuiper belt. The most notable example is the Pluto–Charon binary, but it is estimated that around 11% of KBOs exist in binaries.[88]

1.8 Exploration

Main article: New Horizons

On January 19, 2006, the first spacecraft mission to explore the Kuiper belt, *New Horizons*, was launched. The mission, headed by Alan Stern of the Southwest Research Institute, flew by Pluto on July 14 2015.

Scientists awaited data from the Pan-STARRS survey project to ensure as wide a field of options as possible.[90] The Pan-STARRS project, partially operational since May 2010,[91] will, when fully online, survey the entire sky with four 1.4 gigapixel digital cameras to detect any moving objects, from near-Earth objects to KBOs.[92] To speed up the detection process, the New Horizons team established Ice Hunters, a citizen science project that allowed members of the public to participate in the search for suitable KBO targets;[93][94][95] the project has subsequently been transferred to another site, Ice Investigators,[96] produced by CosmoQuest.[97]

On October 15, 2014, it was revealed that Hubble's search had uncovered three potential targets,[89][98][99][100][101] provisionally designated PT1 ("potential target 1"), PT2 and PT3 by the *New Horizons* team. All are objects with estimated diameters in the 30–55 km range, too small to be seen by ground telescopes, at distances from the Sun of 43–44 AU, which would put the encounters in the 2018–2019 period.[98] The initial estimated probabilities that these objects are reachable within *New Horizons* ' fuel budget are 100%, 7%, and 97%, respectively.[98] All are members of the "cold" (low-inclination, low-eccentricity) classical Kuiper belt, and thus very different from Pluto. PT1 (given the temporary designation "1110113Y" on the HST web site[102]), the most favorably situated object, is magnitude 26.8, 30–45 km in diameter, and will be encountered around January 2019.[103] A course to reach it will require about 35% of *New Horizons* ' available trajectory-adjustment fuel supply. A mission to PT3 was in some ways preferable, in that it is brighter and therefore probably larger than PT1, but the greater fuel requirements to reach it would have left less for maneuvering and unforeseen events.[98] Once sufficient orbital information was provided, the Minor Planet Center gave official designations to the three target KBOs: 2014 MU69 (PT1), 2014 OS393 (PT2), and 2014 PN70 (PT3). By the fall of 2014, a possible fourth target, 2014 MT69, had been eliminated by follow-up observations. PT2 was out of the running before the Pluto flyby.[104][105]

On August 28, 2015, the first target, 2014 MU69, was chosen. Course adjustment will occur in late October and early November for a January 2019 flyby.[106] In order to complete the mission, funding will need to be secured following a senior review of planetary science missions in 2016, with the results of that review to be announced in August or September 2016.[107]

1.9   Extrasolar Kuiper belts

Main article: Debris disc

By 2006, astronomers had resolved dust discs thought to be Kuiper belt-like structures around nine stars other than the Sun. They appear to fall into two categories: wide belts, with radii of over 50 AU, and narrow belts (tentatively like that of the Solar System) with radii of between 20 and 30 AU and relatively sharp boundaries.[108] Beyond this, 15–20% of solar-type stars have an observed infrared excess that is suggestive of massive Kuiper-belt-like structures.[109] Most known debris discs around other stars are fairly young, but the two images on the right, taken by the Hubble Space Telescope in January 2006, are old enough (roughly 300 million years) to have settled into stable configurations. The left image is a "top view" of a wide belt, and the right image is an "edge view" of a narrow belt.[108][110] Computer simulations of dust in the Kuiper belt suggest that when it was younger, it may have resembled the narrow rings seen around younger stars.[111]

1.10   See also

- List of possible dwarf planets

- List of trans-Neptunian objects

1.11   Notes

[1] The literature is inconsistent in the usage of the terms *scattered disc* and *Kuiper belt*. For some, they are distinct populations; for others, the scattered disc is part of the Kuiper belt. Authors may even switch between these two uses in a single publication.[10] Because the International Astronomical Union's Minor Planet Center, the body responsible for cataloguing minor planets in the Solar System, makes the distinction,[11] the current editorial choice for Wikipedia articles on the trans-Neptunian region is to make this distinction as well. This choice means that, on Wikipedia, Eris, the most-massive known trans-Neptunian object, is not part of the Kuiper belt, and this makes Pluto the most-massive Kuiper belt object.

1.12   References

[1] Kuiper belt - oxforddictionaries.com

[2] Alan Stern; Colwell, Joshua E. (1997). "Collisional Erosion in the Primordial Edgeworth-Kuiper Belt and the Generation of the 30–50 AU Kuiper Gap". *The Astrophysical Journal* **490** (2): 879–882. Bibcode:1997ApJ...490..879S. doi:10.1086/304912.

[3] Audrey Delsanti & David Jewitt. "The Solar System Beyond The Planets" (PDF). *Institute for Astronomy, University of Hawaii.* Archived from the original (PDF) on September 25, 2007. Retrieved March 9, 2007.

[4] Krasinsky, G. A.; Pitjeva, E. V.; Vasilyev, M. V.; Yagudina, E. I. (July 2002). "Hidden Mass in the Asteroid Belt". *Icarus* **158** (1): 98–105. Bibcode:2002Icar..158...98K. doi:10.1006/icar.2002.6837.

[5] Johnson, Torrence V.; and Lunine, Jonathan I.; *Saturn's moon Phoebe as a captured body from the outer Solar System*, Nature, Vol. 435, pp. 69–71

[6] Craig B. Agnor & Douglas P. Hamilton (2006). "Neptune's capture of its moon Triton in a binary-planet gravitational encounter" (PDF). *Nature.* Archived from the original (PDF) on June 21, 2007. Retrieved June 20, 2006.

[7] Jewitt, David; Luu, Jane (1993). "Discovery of the candidate Kuiper belt object 1992 QB1". *Nature* **362** (6422): 730. Bibcode:1993Natur.362..730J. doi:10.1038/362730a0.

[8] NEW HORIZONS *The PI's Perspective*

[9] Harold F. Levison; Luke Donnes (2007). "Comet Populations and Cometary Dynamics". In Lucy Ann Adams McFadden; Paul Robert Weissman; Torrence V. Johnson. *Encyclopedia of the Solar System* (2nd ed.). Amsterdam; Boston: Academic Press. pp. 575–588. ISBN 0-12-088589-1.

[10] Weissman and Johnson, 2007, *Encyclopedia of the solar system*, footnote p. 584

[11] IAU: Minor Planet Center (2011-01-03). "List Of Centaurs and Scattered-Disk Objects". Central Bureau for Astronomical Telegrams, Harvard-Smithsonian Center for Astrophysics. Retrieved 2011-01-03.

[12] Gérard FAURE (2004). "Description of the System of Asteroids as of May 20, 2004". Archived from the original on May 29, 2007. Retrieved June 1, 2007.

[13] "What is improper about the term "Kuiper belt"? (or, Why name a thing after a man who didn't believe its existence?)". *International Comet Quarterly.* Retrieved October 24, 2010.

[14] J. K. Davies; J. McFarland; M. E. Bailey; B. G. Marsden; W. I. Ip (2008). "The Early Development of Ideas Concerning the Transneptunian Region". In M. Antonietta Baracci; Hermann Boenhardt; Dale Cruikchank; Alissandro Morbidelli. *The Solar System Beyond Neptune* (PDF). University of Arizona Press. pp. 11–23.

[15] John Davies (2001). *Beyond Pluto: Exploring the outer limits of the solar system.* Cambridge University Press. xii.

[16] Davies, p. 2

[17] David Jewitt. "WHY "KUIPER" BELT?". *University of Hawaii.* Retrieved June 14, 2007.

[18] Davies, p. 14

[19] Rao, M. M. (1964). "Decomposition of Vector Measures" (PDF). *Proceedings of the National Academy of Sciences* **51** (5): 771. Bibcode:1964PNAS...51..771R. doi:10.1073/pnas.51.5.771.

[20] CT Kowal; W Liller; BG Marsden (1977). "The discovery and orbit of /2060/ Chiron". *In: Dynamics of the solar system; Proceedings of the Symposium* (Hale Observatories, Harvard–Smithsonian Center for Astrophysics) **81**: 245. Bibcode:1979IAUS...81..245K.

[21] JV Scotti; DL Rabinowitz; CS Shoemaker; EM Shoemaker; DH Levy; TM King; EF Helin; J Alu; K Lawrence; RH McNaught; L Frederick; D Tholen; BEA Mueller (1992). "1992 AD". *IAU Circ.* **5434**: 1. Bibcode:1992IAUC.5434....1S.

[22] Horner, J.; Evans, N.W.; Bailey, M. E. (2004). "Simulations of the Population of Centaurs I: The Bulk Statistics". *MNRAS* **354** (3): 798–810. arXiv:astro-ph/0407400. Bibcode:2004MNRAS.354..798H. doi:10.1111/j.1365-2966.2004.08240.x.

[23] Davies p. 38

[24] David Jewitt (2002). "From Kuiper Belt Object to Cometary Nucleus: The Missing Ultrared Matter". *The Astronomical Journal* **123** (2): 1039–1049. Bibcode:2002AJ....123.1039J. doi:10.1086/338692.

[25] Oort, J. H. (1950). "The structure of the cloud of comets surrounding the Solar System and a hypothesis concerning its origin". *Bull. Astron. Inst. Neth.* **11**: 91. Bibcode:1950BAN....11...91O.

[26] Davies p. 39

[27] JA Fernández (1980). "On the existence of a comet belt beyond Neptune". *Monthly Notices of the Royal Astronomical Society* **192**: 481. Bibcode:1980MNRAS.192..481F. doi:10.1093/mnras/192.3.481.

[28] M. Duncan; T. Quinn & S. Tremaine (1988). "The origin of short-period comets". *Astrophysical Journal* **328**: L69. Bibcode:1988ApJ...328L.. doi:10.1086/185162.

[29] Davies p. 191

[30] Davies p. 50

[31] Davies p. 51

[32] Davies pp. 52, 54, 56

[33] Davies pp. 57, 62

[34] Davies p. 65

[35] BS Marsden; Jewitt, D.; Marsden, B. G. (1993). "1993 FW". *IAU Circ.* (Minor Planet Center) **5730**: 1. Bibcode:1993IAUC.5730....1L.

[36] Davies p. 199

[37] Clyde Tombaugh, "The Last Word", Letters to the Editor, *Sky & Telescope*, December, 1994, p. 8

[38] Hansen, K. (7 June 2005). "Orbital shuffle for early solar system". *Geotimes*. Retrieved 2007-08-26.

[39] Tsiganis, K.; Gomes, R.; Morbidelli, A.; Levison, H. F. (2005). "Origin of the orbital architecture of the giant planets of the Solar System". *Nature* **435** (7041): 459–461. Bibcode:2005Natur.435..459T. doi:10.1038/nature03539. PMID 15917800.

[40] Levison, H. F.; Morbidelli, A.; Van Laerhoven, C.; Gomes, R. (2008). "Origin of the structure of the Kuiper belt during a dynamical instability in the orbits of Uranus and Neptune". *Icarus* **196** (1): 258–273. arXiv:0712.0553. Bibcode:2008Icar..196..258L. doi:10.1016/j.icarus.2007.11.035.

[41] Thommes, E. W.; Duncan, M. J.; Levison, H. F. (2002). "The Formation of Uranus and Neptune among Jupiter and Saturn". *The Astronomical Journal* **123** (5): 2862. arXiv:astro-ph/0111290. Bibcode:2002AJ....123.2862T. doi:10.1086/339975.

[42] Malhotra, R. (1994). "Nonlinear Resonances in the Solar System". *Physica D* **77**: 289. arXiv:chao-dyn/9406004. Bibcode:1994PhyD...77..28 doi:10.1016/0167-2789(94)90141-4.

[43] Lovett, R. (2010). "Kuiper Belt may be born of collisions". *Nature*. doi:10.1038/news.2010.522.

[44] M. C. De Sanctis; M. T. Capria & A. Coradini (2001). "Thermal Evolution and Differentiation of Edgeworth-Kuiper Belt Objects". *The Astronomical Journal* **121** (5): 2792–2799. Bibcode:2001AJ....121.2792D. doi:10.1086/320385.

[45] "Discovering the Edge of the Solar System". *American Scientists.org*. 2003. Archived from the original on March 5, 2008. Retrieved June 23, 2007.

[46] Michael E. Brown; Margaret Pan (2004). "The Plane of the Kuiper Belt". *The Astronomical Journal* **127** (4): 2418–2423. Bibcode:2004AJ....127.2418B. doi:10.1086/382515.

[47] Jean-Marc Petit; Alessandro Morbidelli; Giovanni B. Valsecchi (1998). "Large Scattered Planetesimals and the Excitation of the Small Body Belts" (PDF). Retrieved June 23, 2007.

[48] Lunine, J. (2003). "The Kuiper Belt" (PDF). Retrieved 2007-06-23.

[49] Jewitt, D. (February 2000). "Classical Kuiper Belt Objects (CKBOs)". Archived from the original on 2007-06-09. Retrieved 2007-06-23.

[50] Murdin, P. (2000). "Cubewano". *The Encyclopedia of Astronomy and Astrophysics*. Bibcode:2000eaa..bookE5403.. doi:10.1888/0333750888 ISBN 0-333-75088-8.

[51] Elliot, J. L.; et al. (2005). "The Deep Ecliptic Survey: A Search for Kuiper Belt Objects and Centaurs. II. Dynamical Classification, the Kuiper Belt Plane, and the Core Population" (PDF). *The Astronomical Journal* **129**: 1117–1162. Bibcode:2005AJ....129.1117E. doi:10.1086/427395.

[52] "Naming of Astronomical Objects: Minor Planets". International Astronomical Union. Retrieved 2008-11-17.

[53] Levison, H. F; Morbidelli, A. (2003). "The formation of the Kuiper belt by the outward transport of bodies during Neptune's migration". *Nature* **426** (6965): 419–421. Bibcode:2003Natur.426..419L. doi:10.1038/nature02120. PMID 14647375.

[54] Morbidelli, A. (2005). "Origin and Dynamical Evolution of Comets and their Reservoirs". arXiv:astro-ph/0512256 [astro-ph].

[55] "List Of Transneptunian Objects". *Minor Planet Center*. Retrieved June 23, 2007.

[56] Chiang; et al. (2003). "Resonance Occupation in the Kuiper Belt: Case Examples of the 5:2 and Trojan Resonances". *The Astronomical Journal* **126** (1): 430–443. arXiv:astro-ph/0301458. Bibcode:2003AJ....126..430C. doi:10.1086/375207.

[57] Wm. Robert Johnston (2007). "Trans-Neptunian Objects". Retrieved June 23, 2007.

[58] Davies p. 104

[59] Davies p. 107

[60] E. I. Chiang & M. E. Brown (1999). "Keck Pencil-Beam Survey For Faint Kuiper Belt Objects" (PDF). Retrieved July 1, 2007.

[61] G.M. Bernstein; D.E. Trilling; R.L. Allen; M.E. Brown; M. Holman & R. Malhotra (2004). "The Size Distribution of Trans-Neptunian Bodies" (PDF). *The Astrophysical Journal* **128** (3): 1364. arXiv:astro-ph/0308467. Bibcode:2004AJ....128.1364B. doi:10.1086/422919.

[62] Michael Brooks (2007). "13 Things that do not make sense". *NewScientistSpace.com*. Retrieved June 23, 2007.

[63] Govert Schilling (2008). "The mystery of Planet X". *New Scientist*. Retrieved February 8, 2008.

[64] Stephen C. Tegler (2007). "Kuiper Belt Objects: Physical Studies". In Lucy-Ann McFadden; et al. *Encyclopedia of the Solar System*. pp. 605–620.

[65] Altwegg, K.; Balsiger, H.; Geiss, J. (1999). "Composition of the Volatile Material in Halley's Coma from In Situ Measurements". *Space Science Reviews* **90**: 3–18. Bibcode:1999SSRv...90....3A. doi:10.1023/A:1005256607402.

[66] David C. Jewitt & Jane Luu (2004). "Crystalline water ice on the Kuiper belt object (50000) Quaoar" (PDF). Archived from the original (PDF) on June 21, 2007. Retrieved June 21, 2007.

[67] Dave Jewitt (2004). "Surfaces of Kuiper Belt Objects". *University of Hawaii*. Archived from the original on June 9, 2007. Retrieved June 21, 2007.

[68] Jewitt, David; Luu, Jane (1998). "Optical-Infrared Spectral Diversity in the Kuiper Belt". *The Astronomical Journal* **115** (4): 1667. Bibcode:1998AJ....115.1667J. doi:10.1086/300299.

[69] Davies p. 118

[70] Jewitt, David C.; Luu, Jane X. (2001). "Colors and Spectra of Kuiper Belt Objects". *The Astronomical Journal* **122** (4): 2099. arXiv:astro-ph/0107277. Bibcode:2001AJ....122.2099J. doi:10.1086/323304.

[71] Brown, R. H.; Cruikshank, DP; Pendleton, Y; Veeder, GJ (1997). "Surface Composition of Kuiper Belt Object 1993SC". *Science* **276** (5314): 937–9. Bibcode:1997Sci...276..937B. doi:10.1126/science.276.5314.937. PMID 9163038.

[72] Brown, Michael E.; Blake, Geoffrey A.; Kessler, Jacqueline E. (2000). "Near-Infrared Spectroscopy of the Bright Kuiper Belt Object 2000 EB173". *The Astrophysical Journal* **543** (2): L163. Bibcode:2000ApJ...543L.163B. doi:10.1086/317277.

[73] Licandro; Oliva; Di MArtino (2001). "NICS-TNG infrared spectroscopy of trans-neptunian objects 2000 EB173 and 2000 WR106". *Astronomy and Astrophysics* **373** (3): L29. arXiv:astro-ph/0105434. Bibcode:2001A&A...373L..29L. doi:10.1051/0004-6361:20010758.

[74] Gladman, Brett; et al. (August 2001). "The structure of the Kuiper belt". *Astronomical Journal* **122** (2): 1051–1066. Bibcode:2001AJ....122.1051G. doi:10.1086/322080.

[75] Lorenzo Iorio (2007). "Dynamical determination of the mass of the Kuiper Belt from motions of the inner planets of the Solar system". *Monthly Notices of the Royal Astronomical Society* **375** (4): 1311–1314. Bibcode:2007MNRAS.tmp...24I. doi:10.1111/j.1365-2966.2006.11384.x.

[76] Bernstein, G. M.; Trilling, D. E.; Allen, R. L.; Brown, K. E.; Holman, M.; Malhotra, R. (2004). "The size distribution of transneptunian bodies". *The Astronomical Journal* **128** (3): 1364–1390. arXiv:astro-ph/0308467. Bibcode:2004AJ....128.1364B. doi:10.1086/422919.

[77] "Hubble Finds Smallest Kuiper Belt Object Ever Seen". HubbleSite. December 2009. Retrieved June 29, 2015.

[78] "List Of Centaurs and Scattered-Disk Objects". *IAU: Minor Planet Center*. Retrieved October 27, 2010.

[79] David Jewitt (2005). "The 1000 km Scale KBOs". *University of Hawaii*. Retrieved July 16, 2006.

[80] Craig B. Agnor & Douglas P. Hamilton (2006). "Neptune's capture of its moon Triton in a binary-planet gravitational encounter" (PDF). *Nature*. Archived from the original (PDF) on June 21, 2007. Retrieved October 29, 2007.

[81] Encrenaz, Thérèse; Kallenbach, R.; Owen, T.; Sotin, C. (2004). *TRITON, PLUTO, CENTAURS, AND TRANS-NEPTUNIAN BODIES. NASA Ames Research Center* (Springer). ISBN 978-1-4020-3362-9. Retrieved June 23, 2007.

[82] Mike Brown (2007). "Dysnomia, the moon of Eris". *CalTech*. Retrieved June 14, 2007.

[83] "Resolution B5 and B6" (PDF). International Astronomical Union. 2006.

[84] "Ixion". *eightplanets.net*. Retrieved June 23, 2007.

[85] John Stansberry; Will Grundy; Mike Brown; Dale Cruikshank; John Spencer; David Trilling; Jean-Luc Margot (2007). "Physical Properties of Kuiper Belt and Centaur Objects: Constraints from Spitzer Space Telescope". arXiv:astro-ph/0702538.

[86] "IAU Draft Definition of Planet". *IAU*. 2006. Retrieved October 26, 2007.

[87] Brown, M. E.; Van Dam, M. A.; Bouchez, A. H.; Le Mignant, D.; Campbell, R. D.; Chin, J. C. Y.; Conrad, A.; Hartman, S. K.; Johansson, E. M.; Lafon, R. E.; Rabinowitz, D. L. Rabinowitz; Stomski, P. J., Jr.; Summers, D. M.; Trujillo, C. A.; Wizinowich, P. L. (2006). "Satellites of the Largest Kuiper Belt Objects" (PDF). *The Astrophysical Journal* **639** (1): L43–L46. arXiv:astro-ph/0510029. Bibcode:2006ApJ...639L..43B. doi:10.1086/501524. Retrieved 2011-10-19.

[88] Agnor, C.B.; Hamilton, D.P. (2006). "Neptune's capture of its moon Triton in a binary-planet gravitational encounter" (PDF). *Nature* **441** (7090): 192–4. Bibcode:2006Natur.441..192A. doi:10.1038/nature04792. PMID 16688170.

[89] Brown, Dwayne; Villard, Ray (October 15, 2014). "RELEASE 14-281 NASA's Hubble Telescope Finds Potential Kuiper Belt Targets for New Horizons Pluto Mission". *NASA*. Retrieved October 16, 2014.

[90] Cal Fussman (2006). "The Man Who Finds Planets". *Discover magazine*. Retrieved August 13, 2007.

[91] Institute for Astronomy, University of Hawai (2010). "PS1 goes Operational and begins Science Mission, May 2010". Retrieved August 30, 2010.

[92] "Pan-Starrs: University of Hawaii". 2005. Retrieved August 13, 2007.

[93] "Ice Hunters web site". Zooniverse.Org. Retrieved July 8, 2011.

[94] "Citizen Scientists: Discover a New Horizons Flyby Target". NASA. Jun 21, 2011. Retrieved August 23, 2011.

[95] Lakdawalla, Emily (June 21, 2011). "The most exciting citizen science project ever (to me, anyway)". The Planetary Society. Retrieved August 31, 2011.

[96] "Ice Investigators". *web site*. CosmoQuest. 2012. Retrieved 2012-05-23.

[97] "Finding Ice". *CosmoQuest web site*. CosmoQuest. 2012-05-20. Retrieved 2012-05-23. External link in |work= (help)

[98] Lakdawalla, Emily (October 15, 2014). "Finally! New Horizons has a second target". *Planetary Society blog*. Planetary Society. Archived from the original on October 15, 2014. Retrieved October 15, 2014.

[99] "NASA's Hubble Telescope Finds Potential Kuiper Belt Targets for New Horizons Pluto Mission". *press release*. Johns Hopkins Applied Physics Laboratory. October 15, 2014. Archived from the original on October 16, 2014. Retrieved October 16, 2014.

[100] Wall, Mike (October 15, 2014). "Hubble Telescope Spots Post-Pluto Targets for New Horizons Probe". Space.com. Archived from the original on October 15, 2014. Retrieved October 15, 2014.

[101] Buie, Marc (October 15, 2014). "New Horizons HST KBO Search Results: Status Report" (PDF). Space Telescope Science Institute. p. 23.

[102] "Hubble to Proceed with Full Search for New Horizons Targets". *HubbleSite news release.* Space Telescope Science Institute. July 1, 2014. Retrieved October 15, 2014.

[103] Stromberg, Joseph (April 14, 2015). "NASA's New Horizons probe is visiting Pluto — and just sent back its first color photos". *Vox.* Retrieved April 14, 2015.

[104] Corey S. Powell (March 29, 2015). "Alan Stern on Pluto's Wonders, New Horizons' Lost Twin, and That Whole "Dwarf Planet" Thing". *Discover.*

[105] http://www.hou.usra.edu/meetings/lpsc2015/pdf/1301.pdf

[106] http://space.io9.com/new-horizons-locks-onto-next-target-lets-explore-the-k-1727298103

[107] Foust, Jeff. "Extended Timetable for Decision on New Horizons Extended Mission". *Space News.* Retrieved 22 July 2015.

[108] Kalas, Paul; Graham, James R.; Clampin, Mark C.; Fitzgerald, Michael P. (2006). "First Scattered Light Images of Debris Disks around HD 53143 and HD 139664". *The Astrophysical Journal* **637**: L57. arXiv:astro-ph/0601488. Bibcode:2006ApJ...637L..57K. doi:10.1086/500305.

[109] Trilling, D. E.; Bryden, G.; Beichman, C. A.; Rieke, G. H.; Su, K. Y. L.; Stansberry, J. A.; Blaylock, M.; Stapelfeldt, K. R.; Beeman, J. W.; Haller, E. E. (February 2008). "Debris Disks around Sun-like Stars". *The Astrophysical Journal* **674** (2): 1086–1105. arXiv:0710.5498. Bibcode:2008ApJ...674.1086T. doi:10.1086/525514.

[110] "Dusty Planetary Disks Around Two Nearby Stars Resemble Our Kuiper Belt". 2006. Retrieved July 1, 2007.

[111] Kuchner, M. J.; Stark, C. C. (2010). "Collisional Grooming Models of the Kuiper Belt Dust Cloud". *The Astronomical Journal* **140** (4): 1007–1019. arXiv:1008.0904. Bibcode:2010AJ....140.1007K. doi:10.1088/0004-6256/140/4/1007.

1.13 External links and data sources

- Dave Jewitt's page @ UCLA
  - The belt's name
- List of short period comets by family
- Kuiper Belt Profile by NASA's Solar System Exploration
- The Kuiper Belt Electronic Newsletter
- Wm. Robert Johnston's TNO page
- Minor Planet Center: Plot of the Outer Solar System, illustrating Kuiper gap
- Website of the International Astronomical Union (debating the status of TNOs)
- XXVIth General Assembly 2006
- nature.com article: diagram displaying inner solar system, Kuiper Belt, and Oort Cloud, taken from Alan Stern, S. (2003). "The evolution of comets in the Oort cloud and Kuiper belt". *Nature* **424** (6949): 639–42. doi:10.1038/nature01725. PMID 12904784.
- SPACE.com: Discovery Hints at a Quadrillion Space Rocks Beyond Neptune (Sara Goudarzi) August 15, 2006 06:13 am ET
- The Outer Solar System Astronomy Cast episode No. 64, includes full transcript.
- The Kuiper belt at 365daysofastronomy.org
- Nine Planets' webpage on the Edgeworth-Kuiper Belt and Oort Cloud
- List of TNOS

Astronomer Gerard Kuiper, after whom the Kuiper belt is named

The array of telescopes atop Mauna Kea, with which the Kuiper belt was discovered

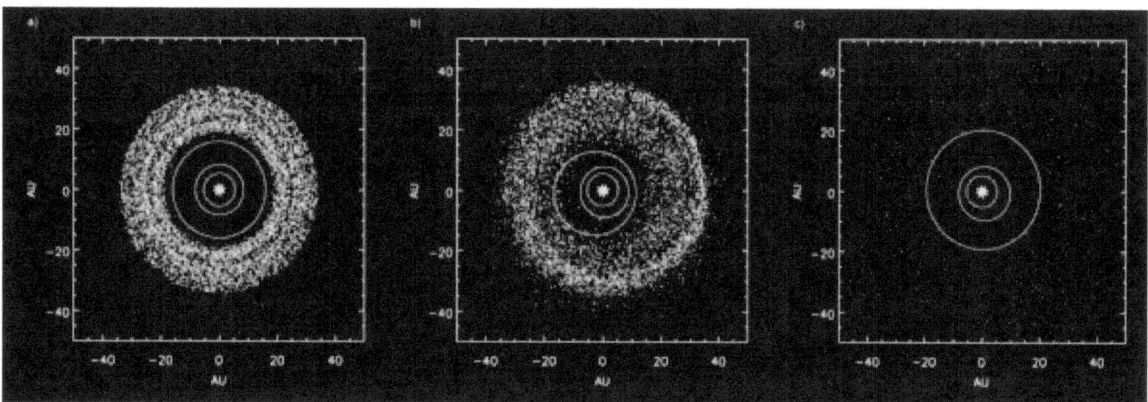

Simulation showing outer planets and Kuiper belt: a) before Jupiter/Saturn 2:1 resonance, b) scattering of Kuiper belt objects into the Solar System after the orbital shift of Neptune, c) after ejection of Kuiper belt bodies by Jupiter

Dust in the Kuiper belt creates a faint infrared disc.

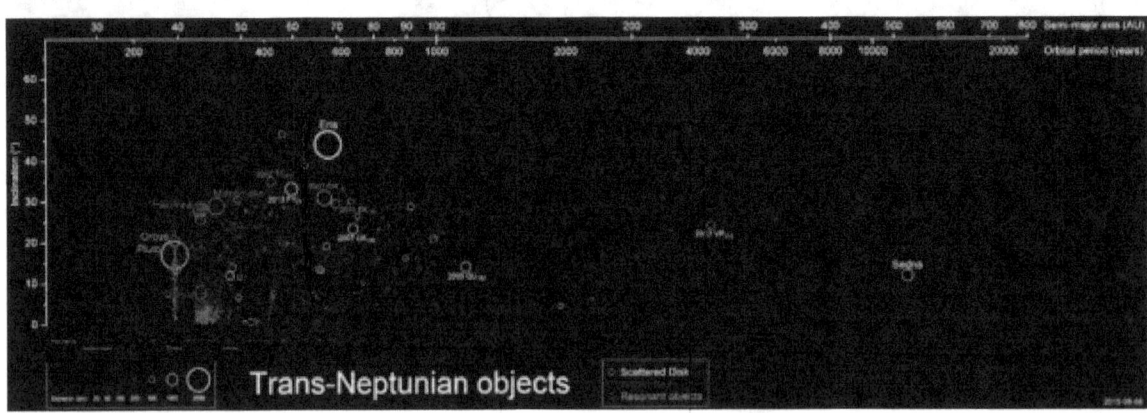

Distribution of cubewanos (blue), Resonant trans-Neptunian objects (red) and scattered objects (grey).

Kuiper belt and orbital resonance

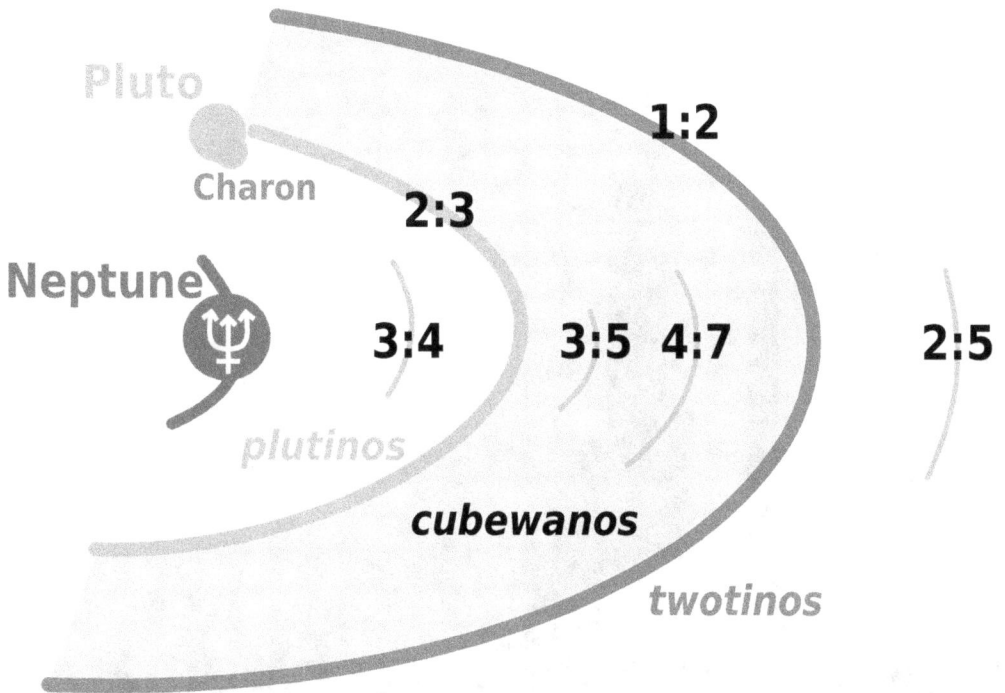

Orbit classification (schematic of semi-major axes)

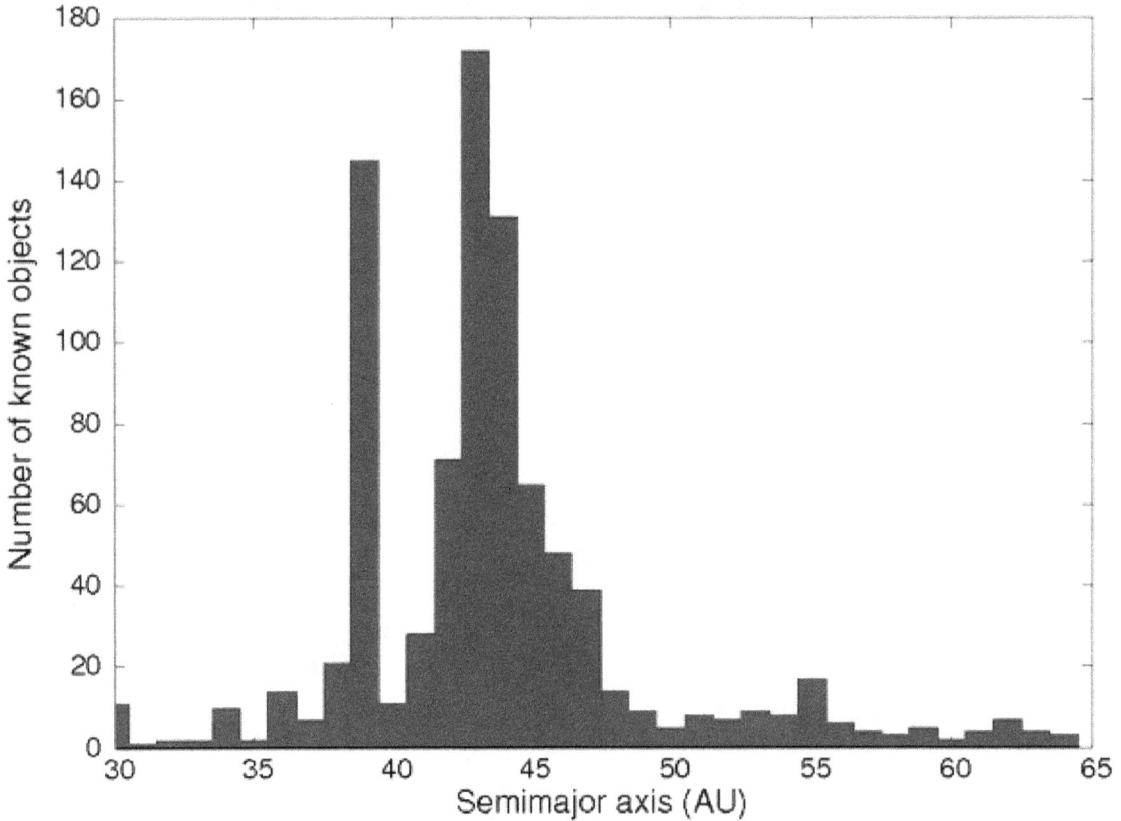

Graph showing the numbers of KBOs for a given distance from the Sun. The plutinos are the "spike" at 39 AU, whereas the classicals are between 42 and 47 AU, the twotinos are at 48 AU, and the 5:2 resonance is at 55 AU.

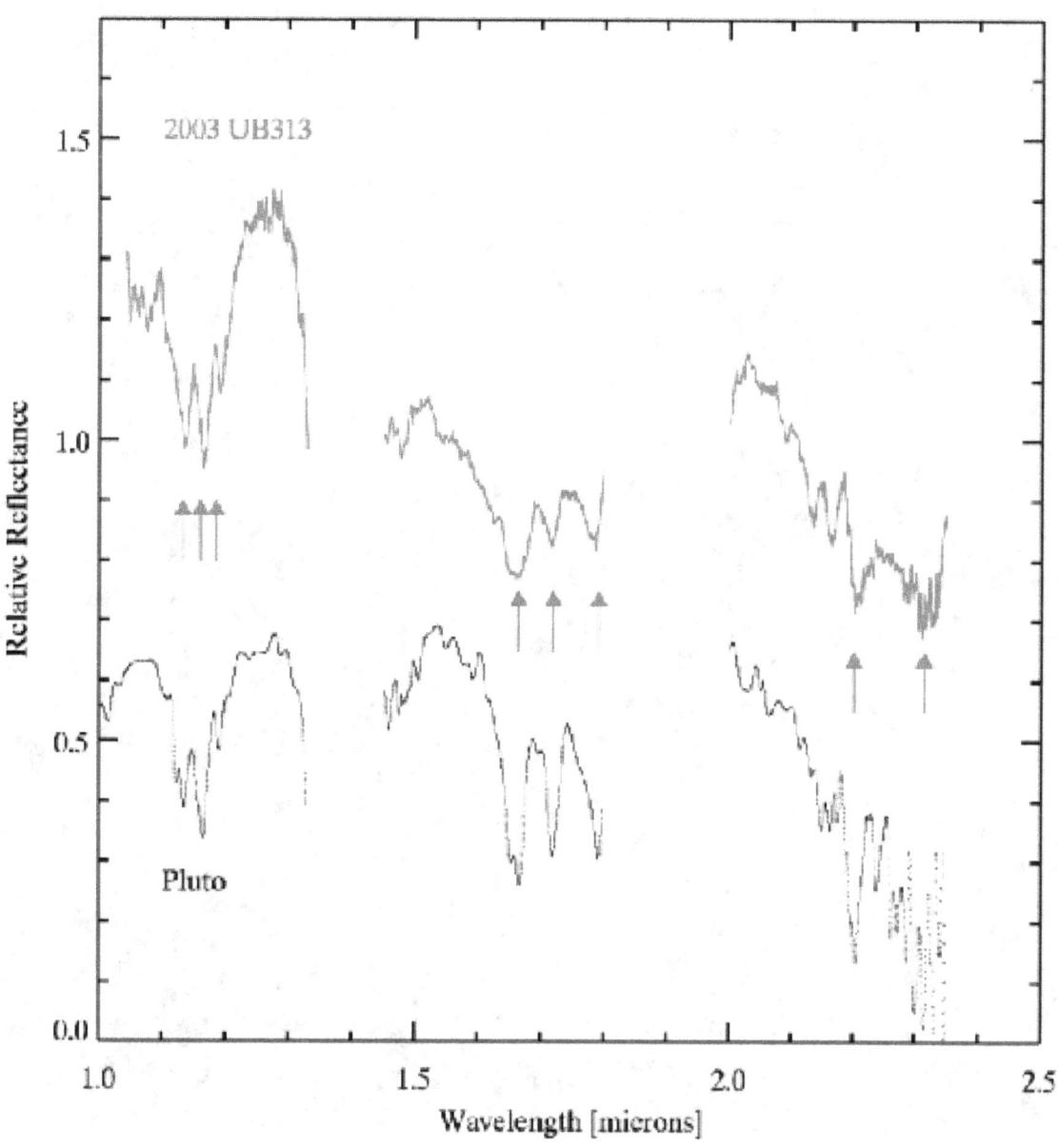

The infrared spectra of both Eris and Pluto, highlighting their common methane absorption lines

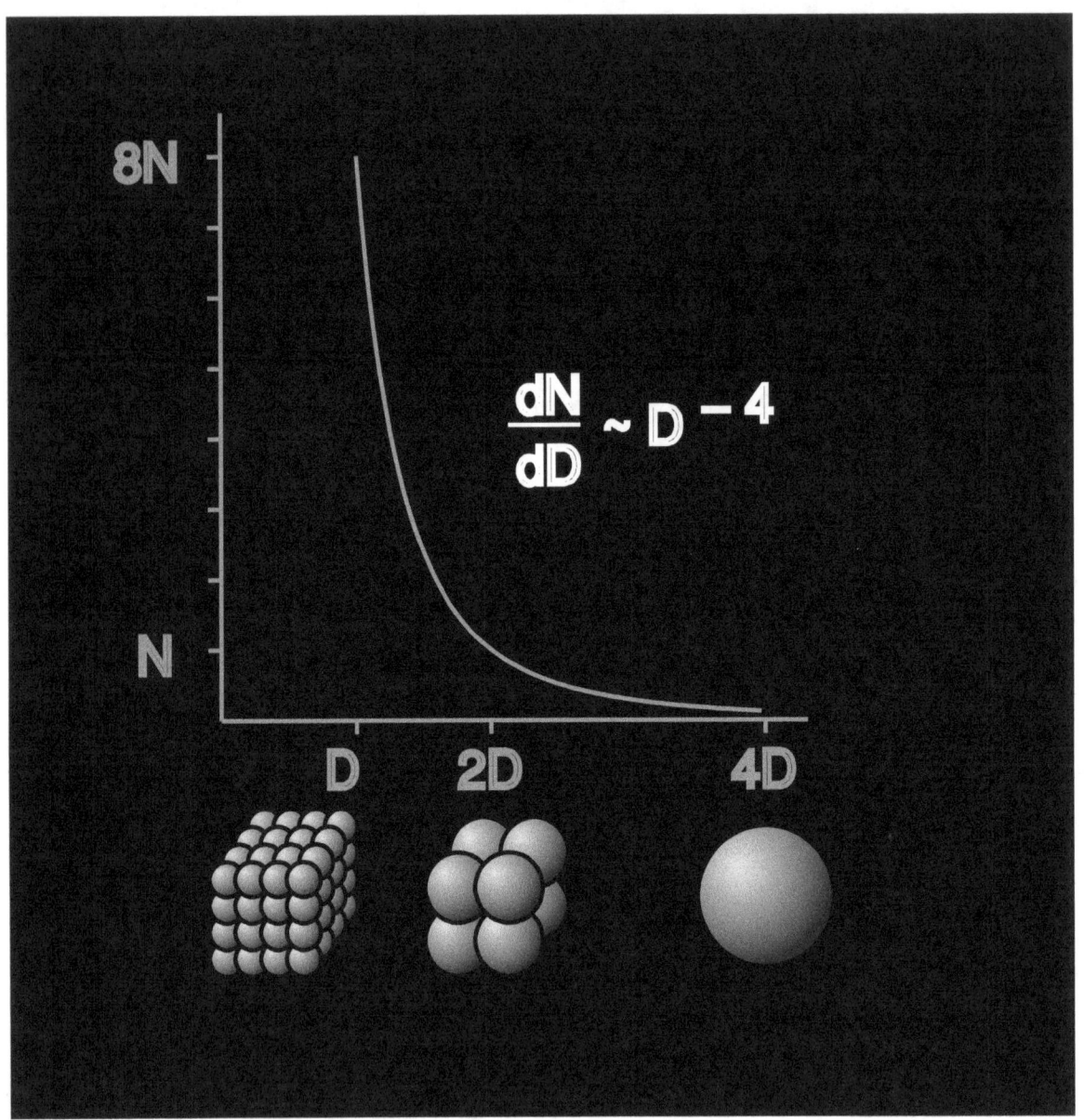

Illustration of the power law.

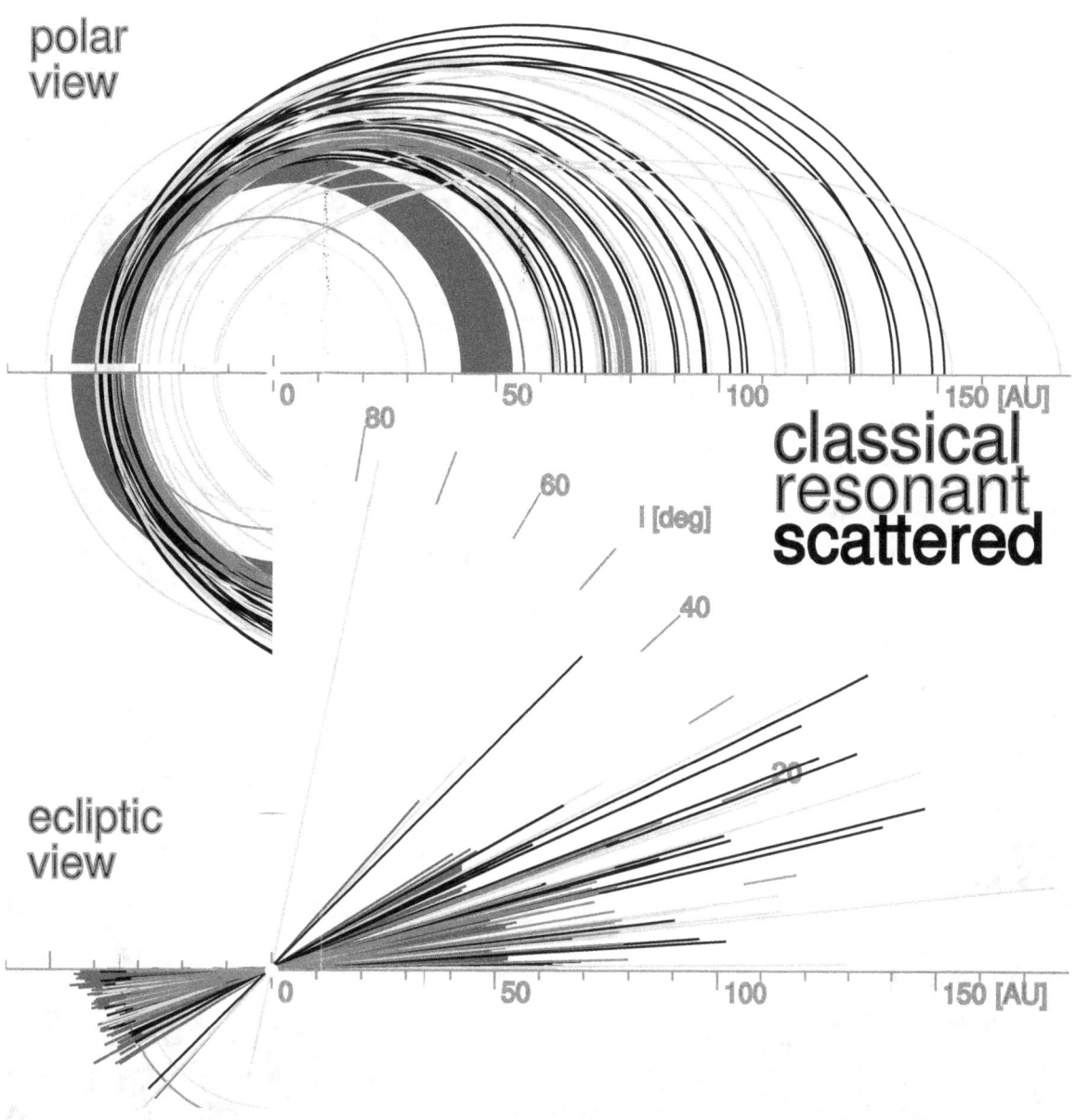

Comparison of the orbits of scattered disc objects (black), classical KBOs (blue), and 2:5 resonant objects (green). Orbits of other KBOs are gray. (Orbital axes have been aligned for comparison.)

Neptune's moon Triton

Kuiper belt object—possible target of New Horizons *spacecraft (artist's concept).*[89]

The KBO 2014 MU69 (green circles), the selected target for the New Horizons Kuiper belt object mission

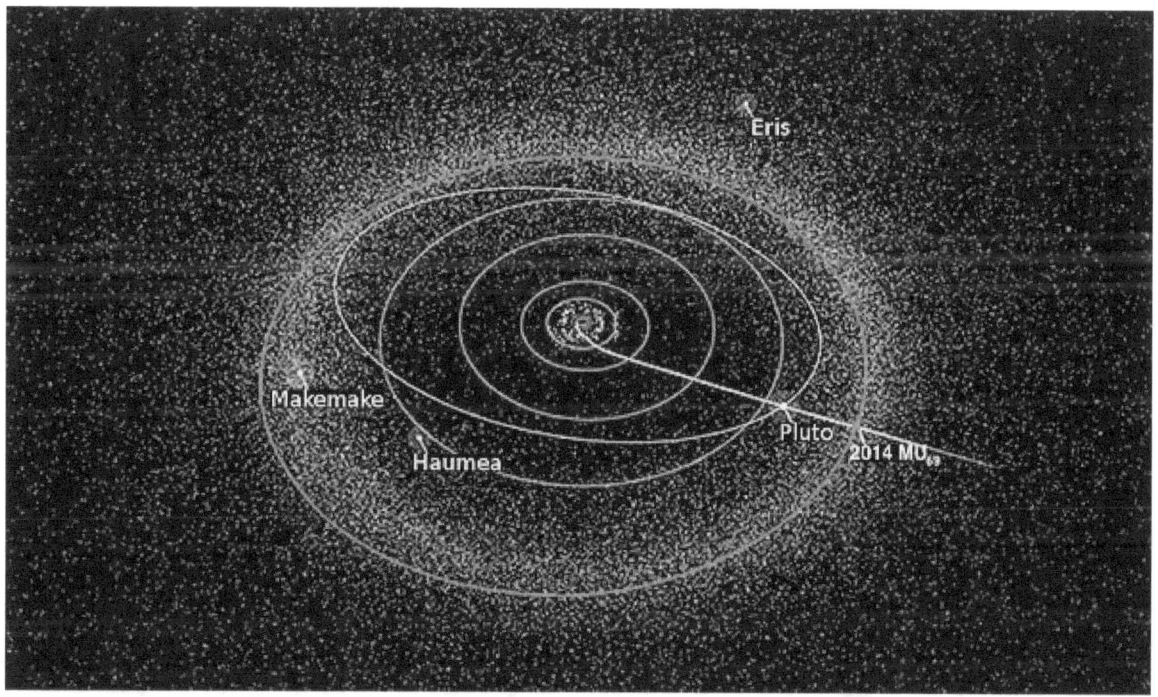

Diagram showing the location of 2014 MU69 and trajectory for rendezvous

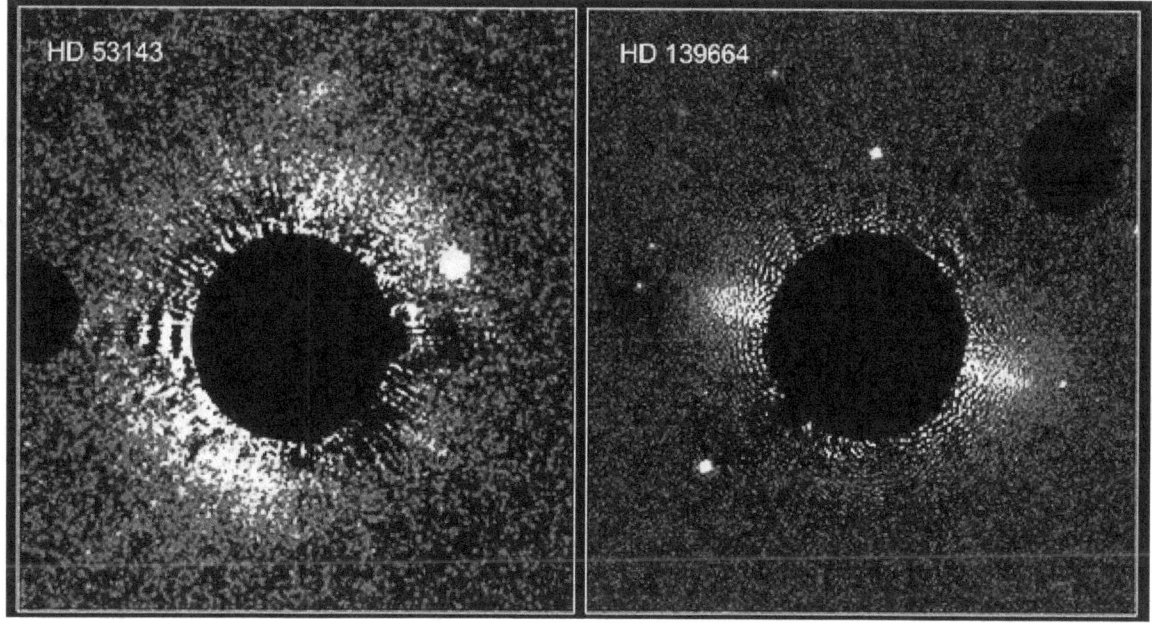

Debris discs around the stars HD 139664 and HD 53143 – black circle from camera hides star to display discs.

Chapter 2

List of the brightest Kuiper belt objects

Main article: Kuiper belt

Since the year 2000, a number of Kuiper belt objects (KBOs) with diameters of between 500 and 1500 km (more than half that of Pluto) have been discovered. 50000 Quaoar, a classical KBO discovered in 2002, is over 1200 km across. Makemake and Haumea, both announced on 29 July 2005, are larger still. Other objects, such as 28978 Ixion (discovered in 2001) and 20000 Varuna (discovered in 2000) measure roughly 500 km across.[1] This has led gradually to the acceptance of Pluto as the largest member of the Kuiper belt.

The brightest known dwarf planets and other KBOs (with absolute magnitudes < 4.0), are:

2.1 See also

- List of trans-Neptunian objects

2.2 References

[1] Audrey Delsanti and David Jewitt. "The Solar System Beyond The Planets" (PDF). *Institute for Astronomy, University of Hawaii.* Archived from the original (PDF) on 2007-01-29. Retrieved 2007-03-09.

[2] John Stansberry, Will Grundy, Mike Brown, Dale Cruikshank, John Spencer, David Trilling, Jean-Luc Margot (2007). "Physical Properties of Kuiper Belt and Centaur Objects: Constraints from Spitzer Space Telescope". arXiv:astro-ph/0702538.

[3] NASA Pluto planetary factsheet accessed on July 1, 2007

[4] "MPEC 2005-O42 : 2005 FY9". Retrieved 2007-06-01.

[5] Wm. Robert Johnston. "(136108) 2003 EL61, S/2005 (2003 EL61) 1, and S/2005 (2003 EL61) 2". Retrieved 2007-07-01.

[6] "MPEC 2005-O36 : 2003 EL61".

[7] Wm. Robert Johnston. "Pluto, Charon, Nix, and Hydra". Retrieved 2007-07-01.

[8] http://arxiv.org/pdf/1107.5309v1.pdf

[9] "MPEC 2003-B27 : 2003 AZ84". *Minor Planet Center.* Retrieved 2007-07-02.

Chapter 3

(19308) 1996 TO66

(19308) 1996 TO$_{66}$ (also written **(19308) 1996 TO66**) is a trans-Neptunian object that was discovered in 1996 by Chadwick Trujillo, David Jewitt and Jane Luu.

3.1 Origin

Main article: Haumea family

Based on their common pattern of IR water-ice absorptions, neutral visible spectrum[7] and the clustering of their orbital elements, the other KBOs (24835) 1995 SM55, (55636) 2002 TX300, (120178) 2003 OP32 and (145453) 2005 RR43 all appear to be collisional fragments broken off of the dwarf planet Haumea.

3.2 References

[1] D. Ragozzine; M. E. Brown (2007-09-04). "Candidate Members and Age Estimate of the Family of Kuiper Belt Object 2003 EL$_{61}$". *The Astronomical Journal* **134** (6): 2160–2167. arXiv:0709.0328. Bibcode:2007AJ....134.2160R. doi:10.1086/522334.

[2] "JPL Small-Body Database Browser: 19308 (1996 TO66)" (2003-10-18 last obs). Retrieved 2013-09-04.

[3] Dan Bruton. "Conversion of Absolute Magnitude to Diameter for Minor Planets". Department of Physics & Astronomy (Stephen F. Austin State University). Archived from the original on 23 March 2010. Retrieved 2009-12-27.

[4] Grundy, W. M. (2004). "Diverse albedos of small trans-neptunian objects". *Icarus* **176**: 22. arXiv:astro-ph/0502229. Bibcode:2005Icar..176..184G doi:10.1016/j.icarus.2005.01.007.

[5] Snodgrass, Carry; Dumas, Hainaut (16 December 2009). "Characterisation of candidate members of (136108) Haumea's family". *The Astrophysical Journal*. arXiv:0912.3171. Bibcode:2010A&A...511A..72S. doi:10.1051/0004-6361/200913031.

[6] Tegler, Stephen C. (2007-02-01). "Kuiper Belt Object Magnitudes and Surface Colors". Archived from the original on 1 September 2006. Retrieved 2006-11-07.

[7] Pinilla-Alonso, N.; Licandro, J.; Gil-Hutton, R.; Brunetto, R. (June 2007). "The water ice rich surface of (145453) 2005 RR43: a case for a carbon-depleted population of TNOs?". *Astronomy and Astrophysics* **468** (1): L25. arXiv:astro-ph/0703098. Bibcode:2007A&A...468L..25P. doi:10.1051/0004-6361:20077294.

3.3 External links

- Orbital simulation from JPL (Java) / Horizons Ephemeris

- First Rotation Period of a Kuiper Belt Object Measured – ESO, 5 November 1998

Chapter 4

(35671) 1998 SN165

(35671) 1998 SN$_{165}$ is a trans-Neptunian object. It was discovered on 23 September 1998, by A. Gleason at Steward Observatory.

It was originally classified as a plutino with a 2:3 mean-motion resonance with Neptune,[6] but further observations have established that it is a *cubewano*—a member of the classical Kuiper belt.[1]

With an estimated size of 393+39
−38 km,[4] (35671) 1998 SN$_{165}$ is a possible dwarf planet.

4.1 Notes

[1] Using the older, larger, Spitzer size estimate of 460 km.

4.2 References

[1] Marc W. Buie (2004-10-10). "Orbit Fit and Astrometric record for 35671". SwRI (Space Science Department). Retrieved 2015-06-25.

[2] "MPEC 2006-X45 : Distant Minor Planets". Minor Planet Center & Tamkin Foundation Computer Network. 2006-12-21. Archived from the original on 28 August 2008. Retrieved 2008-07-18.

[3] "JPL Small-Body Database Browser: 35671 (1998 SN165)" (2004-10-10 last obs). Retrieved 2015-06-25.

[4] TNOs are Cool: A survey of the trans-Neptunian region. X. Analysis of classical Kuiper belt objects from Herschel and Spitzer observations p. 18

[5] John Stansberry; Will Grundy; Mike Brown; Dale Cruikshank; John Spencer; David Trilling; Jean-Luc Margot (2007). "Physical Properties of Kuiper Belt and Centaur Objects: Constraints from Spitzer Space Telescope". arXiv:astro-ph/0702538 [astro-ph].

[6] Hutton, Gil (August 2001). "VR Photometry of Sixteen Kuiper Belt Objects" **152** (2). Icarus. pp. 246–250. Retrieved 2007-10-17.

4.3 External links

- Orbital simulation from JPL (Java) / Ephemeris

Chapter 5

38083 Rhadamanthus

38083 Rhadamanthus is a trans-Neptunian object (TNO). It was discovered in 1999 by the Deep Ecliptic Survey. It was originally thought to be a plutino but no longer is.[1][2]

5.1 Discovery and naming

Rhadamanthus was discovered on 17 April 1999 by the Deep Ecliptic Survey.

Rhadamanthus is named after the Greek mythological figure. The name was announced in the circular of the Minor Planet Center of 24 July 2002, which stated "Rhadamanthus was a son of Zeus and Europa. Because of his just and upright life, after death he was appointed a judge of the dead and the ruler of Elysium, a blissfully beautiful area of the Underworld where those favored by the gods spent their life after death. The name was suggested by E. K. Elliot."[6]

5.2 References

[1] Marc W. Buie (7 June 2008). "Orbit Fit and Astrometric record for 38083". SwRI (Space Science Department). Retrieved 2015-10-04.

[2] "MPEC 2006-X45 : Distant Minor Planets". Minor Planet Center & Tamkin Foundation Computer Network. 21 December 2006. Archived from the original on 28 August 2008. Retrieved 2008-07-18.(older provisional Plutino listing)

[3] "JPL Small-Body Database Browser: 38083 Rhadamanthus (1999 HX11)" (2008-06-07 last obs). Retrieved 2008-07-17.

[4] Rhadamanthus

[5] "ABSOLUTE MAGNITUDE (H)". NASA/JPL. Retrieved 2015-09-23.

[6] "Minor Planet Circulars/Minor Planets and Comets, MPC-46112" (PDF). Minor Planet Center, Smithsonian Astrophysical Observatory, Cambridge, MA 02138, U.S.A. Retrieved 2015-04-19.

5.3 External links

- Orbital simulation from JPL (Java)

- Ephemeris

Chapter 6

(15836) 1995 DA2

(15836) 1995 DA$_2$, also written as **(15836) 1995 DA2**, is a trans-Neptunian object. It was discovered on February 24, 1995, by David C. Jewitt and Jane X. Luu at the Mauna Kea Observatory, Hawaii.

6.1 Resonance

It is in a 3:4 resonance with the planet Neptune.[1][2] The Neptune 3:4 mean-motion resonance keeps the object more than 8 AU from Neptune over a 14000-year period.[4][5]

6.2 References

[1] Marc W. Buie (7 February 2002). "Orbit Fit and Astrometric record for 15836". SwRI (Space Science Department). Retrieved 2009-01-29.

[2] "MPEC 2009-A63 :Distant Minor Planets (2009 JAN. 29.0 TT)". Minor Planet Center. 13 January 2009. Retrieved 2009-03-01.

[3] Wm. Robert Johnston (22 August 2008). "List of Known Trans-Neptunian Objects". Johnston's Archive. Archived from the original on 13 February 2009. Retrieved 2009-01-29.

[4] "MPEC 2001-D23". Minor Planet Center. 21 February 2001. Retrieved 2009-01-30.

[5] "MPEC 1996-A11: 1995 DA2". Minor Planet Center. 6 January 1996. Retrieved 2009-01-30.

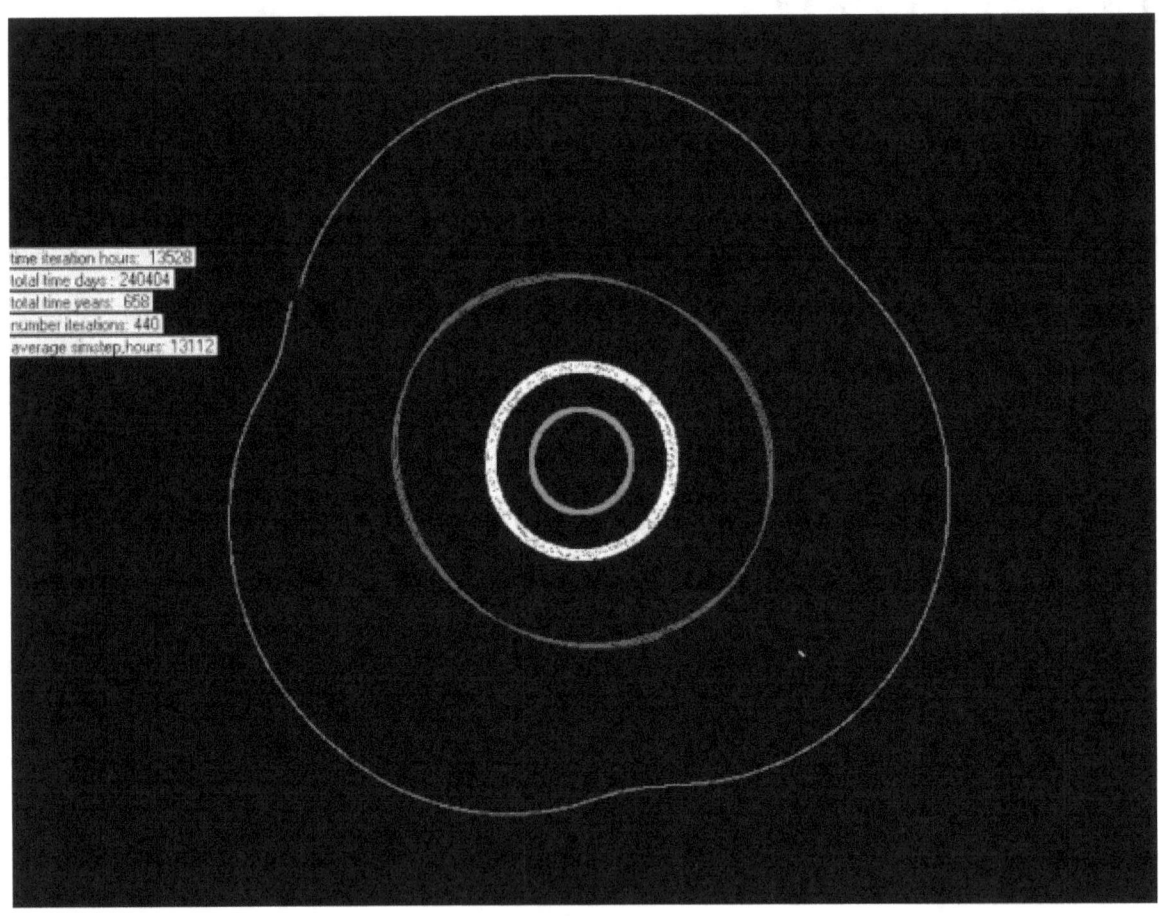

The 3:4 resonance libration of 1995 DA₂. Neptune is the white (stationary) dot at 5 o'clock. Uranus is blue, Saturn yellow, and Jupiter red.

Chapter 7

(15883) 1997 CR29

(15883) 1997 CR$_{29}$, also written as **(15883) 1997 CR29**, is a trans-Neptunian object. It was discovered on February 3, 1997, by Chad Trujillo, Jun Chen, and David C. Jewitt at the Mauna Kea Observatory, Hawaii.

7.1 References

- http://www.minorplanetcenter.org/iau/lists/TNOs.html
- http://www.johnstonsarchive.net/astro/tnoslist.html

Chapter 8

(79978) 1999 CC158

(79978) 1999 CC$_{158}$, also written as **(79978) 1999 CC158**, is a trans-Neptunian object orbiting in the Kuiper belt of the Solar System.[1] It was discovered on 15 February 1999 at the Mauna Kea Observatory, Hawaii. It has a 5:12 resonance with Neptune.[2]

8.1 References

[1] "List Of Centaurs and Scattered-Disk Objects". Minor Planet Center. Retrieved 2009-01-29.

[2] Marc W. Buie. "Orbit Fit and Astrometric record for 79978". SwRI (Space Science Department). Retrieved 2009-01-29. 2007-11-09 using 25 observations

[3] Wm. Robert Johnston (22 August 2008). "List of Known Trans-Neptunian Objects". Johnston's Archive. Archived from the original on 13 February 2009. Retrieved 2009-01-29.

Chapter 9

(182294) 2001 KU76

(182294) 2001 KU$_{76}$, provisionally known as 2001 KU76, is a trans-Neptunian object (TNO) that has a 6:11 resonance with Neptune.[2] This is the same resonance that dwarf planet Makemake is either near or in.[5]

It will come to perihelion in 2021.[1]

Assuming a generic TNO albedo of 0.09, it is about 211 km in diameter.[4] The assumed diameter of this object makes it a possible dwarf planet.[6]

9.1 Resonance

Simulations by Lykawka in 2007 show that (182294) 2001 KU$_{76}$ is librating in the 11:6 resonance with Neptune.[2][3] This is the same resonance that dwarf planet Makemake is either near or in.[5] Both objects have a semi-major axis of 45 AU and an orbital period of about 302 years.

It has been observed 29 times over 6 years and has an orbit quality code of 4.[1]

9.2 References

[1] "JPL Small-Body Database Browser: 182294 (2001 KU76)" (2008-05-03 last obs). Retrieved 2009-02-06.

[2] Lykawka, Patryk Sofia; Mukai, Tadashi (July 2007). "Dynamical classification of trans-neptunian objects: Probing their origin, evolution, and interrelation". *Icarus* 189 (1): 213–232. Bibcode:2007Icar..189..213L. doi:10.1016/j.icarus.2007.01.001.

[3] Buie, Marc W.. "Orbit Fit and Astrometric record for 182294" (2008-05-03 using 29 observations). SwRI (Space Science Department). Retrieved 2009-02-06.

[4] Johnston, Wm. Robert (22 August 2008). "List of Known Trans-Neptunian Objects". Johnston's Archive. Retrieved 2009-02-06.

[5] Tony Dunn (Author of Gravity Simulator). "Possible resonances of Eris (2003 UB$_{313}$) and Makemake (2005 FY$_9$)". Gravity Simulator. Retrieved 2009-02-06.

[6] Brown, Michael E.. "How many dwarf planets are there in the outer solar system? (updates daily)". California Institute of Technology. Retrieved 2012-09-04.

9.3 External links

- Orbital simulation from JPL (Java) / Horizons Ephemeris

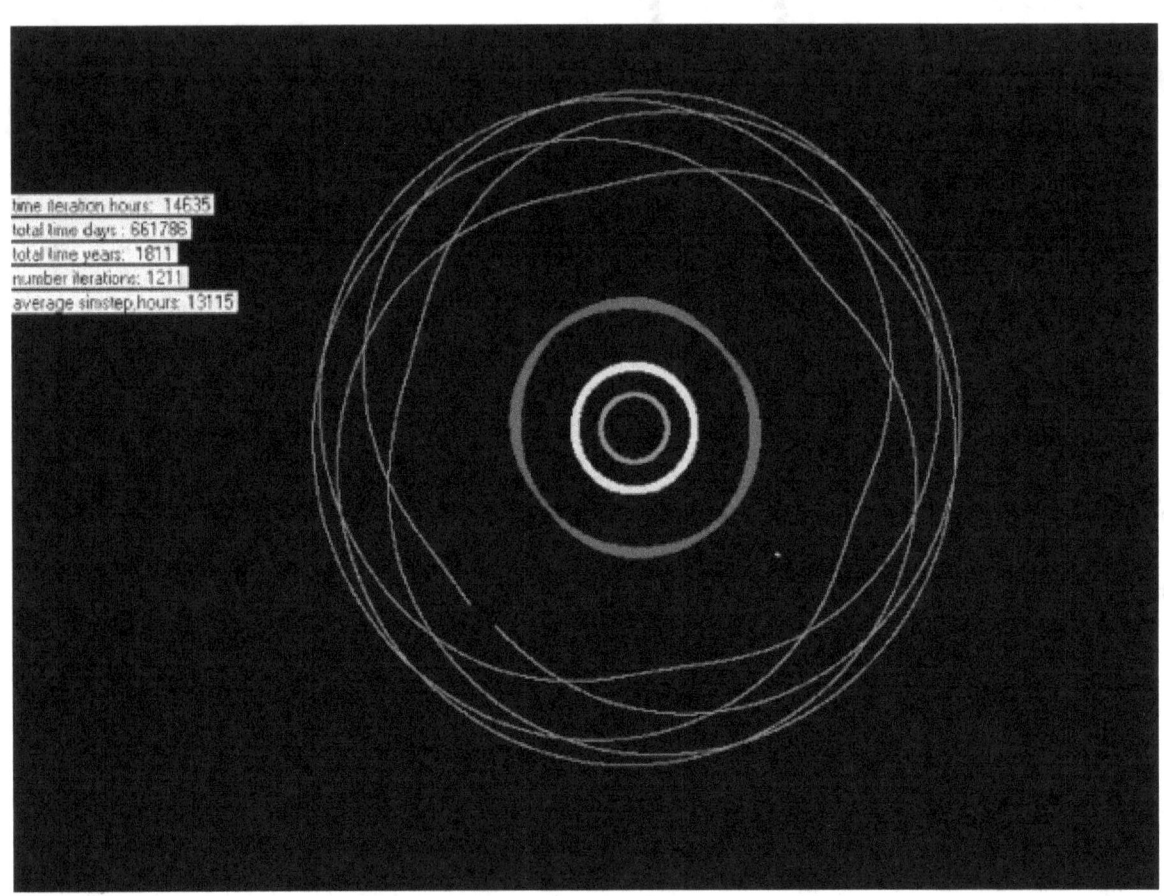

The libration of 2001 KU₇₆. Neptune is the white (stationary) dot at 5 o'clock. Uranus is blue, Saturn yellow, and Jupiter red.

Chapter 10

2003 LA7

2003 LA$_7$, also written as **2003 LA7**, is a resonant trans-Neptunian object that goes around the Sun once for every four times that Neptune goes around. Another possible fourtino is 2011 UP411.

2003 LA$_7$ is in a 1:4 resonance with the planet Neptune.[1][2] For every one orbit that a it makes, Neptune orbits 4 times.

It is currently 43 AU from the Sun,[5] and will come to perihelion around 2041.[3]

Assuming a generic TNO albedo of 0.09, it is about 231 km in diameter.[4]

It has been observed 14 times over 4 oppositions.[3]

10.1 See also

- (119979) 2002 WC19 (a twotino)
- (136120) 2003 LG7 ("threetino")

10.2 References

[1] "MPEC 2009-C70 :Distant Minor Planets (2009 FEB. 28.0 TT)". Minor Planet Center. 2009-02-10. Retrieved 2009-03-14.

[2] Marc W. Buie. "Orbit Fit and Astrometric record for 03LA7" (last observation: 2008-03-12 using 17 of 18 observations). SwRI (Space Science Department). Retrieved 2014-10-13.

[3] "JPL Small-Body Database Browser: (2003 LA7)" (last observation: 2007-04-21). Retrieved 2009-03-14.

[4] Wm. Robert Johnston (22 August 2008). "List of Known Trans-Neptunian Objects". Johnston's Archive. Archived from the original on 13 February 2009. Retrieved 2009-03-14.

[5] "AstDys 2003LA7 Ephemerides". Department of Mathematics, University of Pisa, Italy. Archived from the original on 2009-05-14. Retrieved 2009-03-19.

10.3 External links

- Orbital simulation from JPL (Java) / Horizons Ephemeris

Chapter 11

(136120) 2003 LG7

(136120) 2003 LG$_7$, also written as **2003 LG7**, is a trans-Neptunian object that resides in the Kuiper belt. It was discovered on June 1, 2003 by Marc W. Buie. It is in a 1:3 orbital resonance with Neptune,[2][3] which means that for every one orbit that it makes, Neptune orbits 3 times.

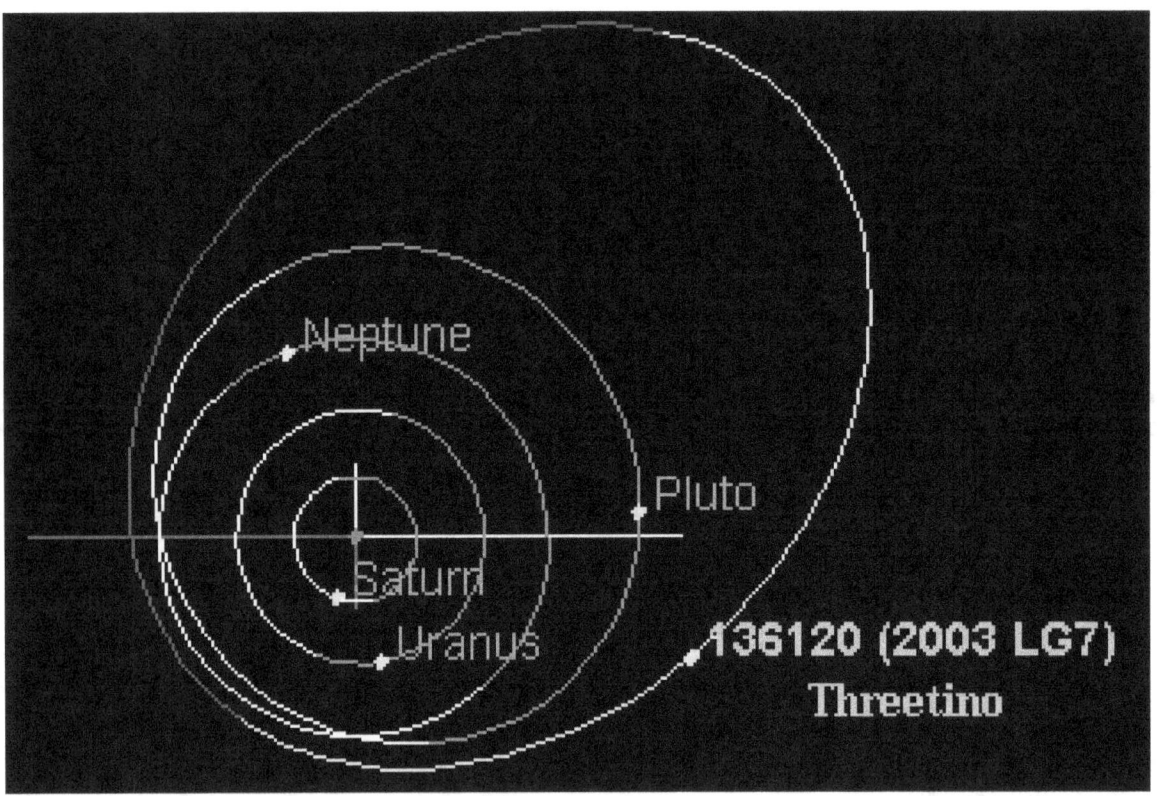

The elongated orbit of "threetino" 2003 LG$_7$ compared to Pluto and Neptune.

11.1   See also

- (119979) 2002 WC19 (a twotino)
- 2003 LA7 ("fourtino")

11.2 References

[1] List Of Centaurs and Scattered-Disk Objects

[2] Marc W. Buie (2006-06-22). "Orbit Fit and Astrometric record for 136120". SwRI (Space Science Department). Retrieved 2013-02-25.

[3] "MPEC 2009-C70 :Distant Minor Planets (2009 FEB. 28.0 TT)". Minor Planet Center. 2009-02-10. Retrieved 2009-03-07.

[4] AstDyS: (136120) 2003LG7

[5] List of known trans-Neptunian objects

11.3 External links

- Orbital simulation from JPL (Java) / Ephemeris

Chapter 12

2010 FX86

2010 FX$_{86}$, also written **2010 FX86**, is a relatively bright classical Kuiper belt object[3][5] with an absolute magnitude of 4.3.[1] It is estimated to be about 600 kilometres (370 mi) in diameter.[4] Astronomer Mike Brown lists it as likely a dwarf planet.[4]

12.1 References

[1] "MPEC 2010-G57 : 2010 FX86". Minorplanetcenter.org. Retrieved 2014-06-13.

[2] "IAU Minor Planet Center". Minorplanetcenter.net. 2010-03-17. Retrieved 2014-06-13.

[3] Alan Chamberlin. "JPL Small-Body Database Browser". Ssd.jpl.nasa.gov. Retrieved 2014-06-13.

[4] Michael E. Brown. "How many dwarf planets are there in the outer solar system? (updates daily)". California Institute of Technology. Retrieved 2011-08-25.

[5] "MPEC 2011-F17 : 2010 FX86". Minorplanetcenter.org. Retrieved 2014-06-13.

Chapter 13

2010 RF43

2010 RF$_{43}$, also written **2010 RF43**, is a trans-Neptunian object with an absolute magnitude of 4.1.[3] It was discovered in 2010 by S. D. Benecchi at the Las Campanas Observatory in Chile.[1] 2010 RF$_{43}$ is currently classified as a scattered disc object.[3][4] Astronomer Mike Brown lists it as highly likely a dwarf planet.[6]

13.1 References

[1] "MPEC 2011-U09 : 2010 RF43". Minorplanetcenter.net. Retrieved 2014-06-13.

[2] (2010 RF43). "Small Solar System Body (2010 RF43)". Comets-asteroids.findthedata.org. Retrieved 2014-06-13.

[3] Alan Chamberlin. "JPL Small-Body Database Browser". Ssd.jpl.nasa.gov. Retrieved 2014-06-13.

[4] Marc W. Buie. "Orbit Fit and Astrometric record for 10RF43" (2013-02-13 using 34 of 36 observations). SwRI (Space Science Department). Retrieved 2013-02-13.

[5] "Absolute Magnitude (H)". NASA / JPL. Retrieved 2010-02-13.

[6] Michael E. Brown. "How many dwarf planets are there in the outer solar system? (updates daily)". California Institute of Technology. Retrieved 2011-08-25.

Chapter 14

2013 FZ27

2013 FZ$_{27}$, also written **2013 FZ27**, is a trans-Neptunian object that, as of 2014, is located near the edge of the Kuiper belt.[3] Its discovery was announced on 2 April 2014.[1] It has an absolute magnitude (H) of 4.0,[3] which makes it likely to be a dwarf planet. Assuming an albedo of 0.15, it would be approximately 500 kilometres (310 mi) in diameter.[4]

2013 FZ$_{27}$ will come to perihelion in September 2090,[lower-alpha 3] at a distance of 37.98AU.[3] As of 2014, it is 49 AU from the Sun and has an apparent magnitude of 21.1.[1]

First detected on 16 March 2013, it had an observation arc of about one year when announced. It came to opposition in late February 2014. Four precovery images, by Pan-STARRS from 21 February 2013, were quickly located.[2] Eight more precovery images, by Pan-STARRS from January and February 2011, have been located, extending the observation arc to 1151 days.[2] Later, three precovery observations by the Sloan Digital Sky Survey in February 2001 were also found, giving it a well-defined 13-year (4782 day) observation arc.

The sednoid 2012 VP113 and the scattered-disc object 2013 FY27 were discovered by the same survey as 2013 FZ$_{27}$ and were announced a few days before.

14.1 Notes

[1] Assuming an albedo of 0.4

[2] Assuming an albedo of 0.08

[3] The 1-sigma uncertainty in the year of perihelion passage is a bit more than a week.[3]

14.2 References

[1] "MPEC 2014-G07 : 2013 FZ27". IAU Minor Planet Center. 2014-04-02. Retrieved 2014-04-02. (K13F27Z)

[2] "2013 FZ27 Orbit" (arc=4782 days over 4 oppositions). IAU Minor Planet Center. Retrieved 2015-05-04.

[3] "JPL Small-Body Database Browser: (2013 FZ27)" (last observation: 2014-03-26; arc: 13.09 years). Jet Propulsion Laboratory. Retrieved 2015-04-13.

[4] Dan Bruton. "Conversion of Absolute Magnitude to Diameter for Minor Planets". Department of Physics & Astronomy (Stephen F. Austin State University). Retrieved 2014-04-06.

[5] Brown, Michael E.. "How many dwarf planets are there in the outer solar system? (updates daily)". California Institute of Technology. Retrieved 2014-04-18.

14.3   External links

- Orbital simulation from JPL (Java) / Horizons Ephemeris

Chapter 15

2014 MT69

2014 MT$_{69}$ (formerly labeled **0720090F** in the context of the Hubble Space Telescope, and **7** in the context of the *New Horizons* mission) is a Kuiper belt object (KBO) and formerly a potential flyby target for the *New Horizons* probe.[3]

15.1   Discovery and naming

2014 MT$_{69}$ was discovered with the help of the Hubble Space Telescope (HST)[7] because the object has a magnitude of 27.3, which is too faint to be observed by ground-based telescopes. Preliminary observations by the HST searching for KBO flyby targets for the *New Horizons* probe started in June 2014, and more intensive observations continued in July and August.[8][9] 2014 MT$_{69}$ was first discovered in observations on June 24, 2014 during the preliminary observations, but it was designated 0720090F at the time,[2] nicknamed "7" for short.[3][4] Its existence as a potential target of the *New Horizons* probe was revealed by NASA in October 2014,[4] but the official name 2014 MT$_{69}$ was not assigned by the Minor Planet Center (MPC) until March 2015 after better orbit information was available.[3] The parameters of the orbit have the extremely large uncertainty of 9 because follow-up observations after discovery eliminated 2014 MT$_{69}$ as a potential target of the *New Horizons* probe, and no further follow-up observations were made.[4]

15.2   Potential targets of the *New Horizons* mission

Having completed its flyby of Pluto, the *New Horizons* space probe will be maneuvered for a flyby of at least one Kuiper belt object (KBO). There were several potential targets considered for the first such flyby. Potential target 2014 MT$_{69}$ has a diameter between 27–92 km (17–57 mi), which is smaller than the other potential targets of the *New Horizons* probe. A potential encounter with the probe initially looked more feasible for 2014 MT$_{69}$ than for 2014 MU69, but follow-up observations eventually ruled out 2014 MT$_{69}$ as a potential target.[3][4] The potential targets for the *New Horizons* probe were PT1 and PT3, the KBOs 2014 MU$_{69}$ and 2014 PN70, and the probe has sufficient fuel to maneuver to either PT1 or PT3. Potential target PT2, the KBO 2014 OS393, is no longer under consideration as a potential target.[10]

On 28 August 2015, the *New Horizons* team announced the selection of 2014 MU$_{69}$ as the next flyby target.[11]

15.3   References

[1]   "2014 MT69". Minor Planet Center. 2014-08-03.

[2]   "Hubble Survey Finds Two Kuiper Belt Objects to Support New Horizons Mission". *HubbleSite news release*. Space Telescope Science Institute. July 1, 2014.

[3]   Zangari, Amanda (March 28, 2015). "Postcards from Pluto". Tumblr.

[4] Buie, Marc (October 15, 2014). "New Horizons HST KBO Search Results: Status Report" (PDF). Space Telescope Science Institute. p. 23.

[5] "JPL Small-Body Database Browser: (2014 MT69)" (2014-08-03 last obs; arc: 40 days). Retrieved 9 May 2015.

[6] "ABSOLUTE MAGNITUDE (H)".

[7] J. R. Spencer, M. W. Buie; et al. (2015). "The Successful Search for a Post-Pluto KBO Flyby Target for New Horizons Using the Hubble Space Telescope" (PDF). *European Planetary Science Congress (EPSC) Abstract* (Copernicus Office).

[8] "Hubble to Proceed with Full Search for New Horizons Targets". *HubbleSite news release.* Space Telescope Science Institute. July 1, 2014.

[9] Schmidt, Klaus (2 July 2014). "*Hubble* to Proceed with Full Search for *New Horizons* Targets". *International Space Fellowship.*

[10] Powell, Corey S. (March 29, 2015). "Alan Stern on Pluto's Wonders, New Horizons' Lost Twin, and That Whole "Dwarf Planet" Thing". *Discover.*

[11] Cofield, Calla (28 August 2015). "Beyond Pluto: 2nd Target Chosen for New Horizons Probe". *Space.com.*

Chapter 16

2014 MU69

2014 MU$_{69}$ (initially called **PT1** and **1110113Y** by the *New Horizons* and Hubble teams, respectively) is a classical Kuiper belt object and the selected target for the *New Horizons* probe to fly by.[4] In August 2015, 2014 MU$_{69}$ was selected for a flyby on 1 January 2019.[7] 2014 MU$_{69}$ is estimated to be 45 kilometres (28 mi) in diameter.[7] After carrying out four course changes on the 22, 25, 28 of October and the 4 of November, *New Horizons* is now heading on a journey toward 2014 MU$_{69}$.[8][9]

16.1   History

16.1.1   Discovery

2014 MU$_{69}$ was discovered on 26 June 2014 by the Hubble Space Telescope during a preliminary survey to find a suitable Kuiper belt object for the *New Horizons* probe to flyby. The discovery required use of the Hubble Space Telescope, because with an apparent magnitude of 26 it is too faint for all but the most powerful telescopes. Hubble is also capable of very precise astrometry and hence a reliable orbit determination.[10][11]

16.1.2   Designation

When 2014 MU$_{69}$ was first observed, it was labelled 1110113Y,[12] and nicknamed "11", for short.[4][3] Its existence as a potential target of the *New Horizons* probe was announced by NASA in October 2014[13][14] and it was designated PT1 ("Potential Target 1"). Its official designation, 2014 MU$_{69}$, was assigned by the Minor Planet Center in March 2015 after sufficient orbital information was gathered.[4]

16.2   Characteristics

Based on its brightness and distance, 2014 MU$_{69}$ is estimated to have a diameter of 30–45 km (20–30 mi).[2] Its orbital period is about 293 years and it has a low inclination and low eccentricity.[15] This unexcited orbit means that it is a cold classical Kuiper belt object which likely has not undergone significant perturbations.[2] Observations in May and July 2015 have greatly reduced the uncertainties in the orbit but have not yet been added to the MPC and JPL databases.[10]

16.3   Exploration

New Horizons probe has completed its flyby of Pluto and has been maneuvered for a flyby of 2014 MU$_{69}$ on 31 December 2018 or 1 January 2019, at which point they will be 43.4 AU from the Sun in the constellation Sagittarius.[16][17][18][19]

16.4 Gallery

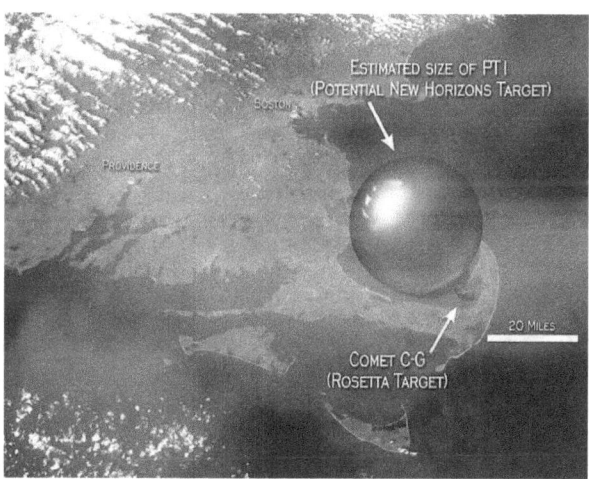

Size of 2014 MU$_{69}$ (PT1) compared to the Eastern Seaboard of the United States and *Rosetta*'s target, comet Churyumov–Gerasimenko.

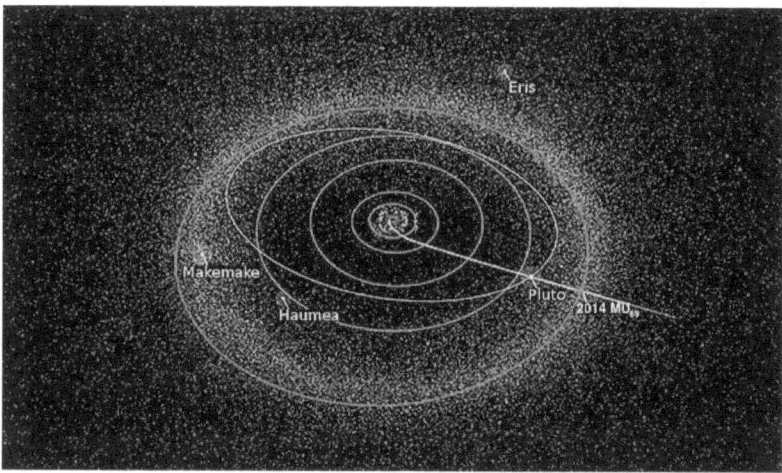

New Horizons trajectory and the orbits of Pluto and 2014 MU$_{69}$.

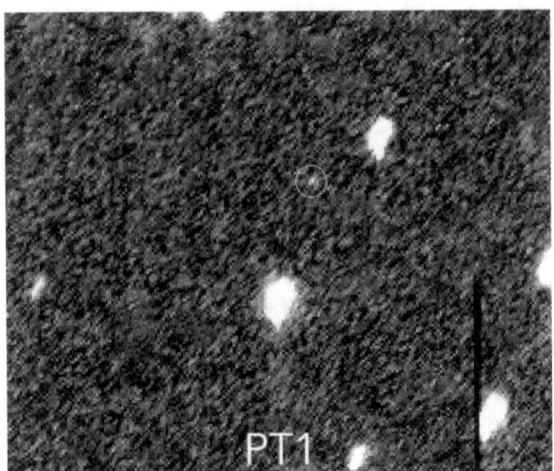

The five discovery images of 2014 MU$_{69}$, shown separately.

16.5   See also

- List of minor planets targeted for spacecraft visitation

16.6   References

[1] "2014 MU69 Orbit". Minor Planet Center. 22 October 2014.

[2] Lakdawalla, Emily (15 October 2014). "Finally! New Horizons has a second target". *Planetary Society blog*. Planetary Society. Archived from the original on 15 October 2014.

[3] Buie, Marc (15 October 2014). "New Horizons HST KBO Search Results: Status Report" (PDF). Space Telescope Science Institute. p. 23.

[4] Talbert, Tricia (28 August 2015). "NASA's New Horizons Team Selects Potential Kuiper Belt Flyby Target". *NASA*. Retrieved 4 September 2015.

[5] "JPL Small-Body Database Browser: (2014 MU69)" (22 October 2014 last obs; arc: 118 days). Retrieved 29 August 2015.

[6] Stern, Alan (August 2015). "OPAG: We Did It!" (PDF). *Presentation to the Outer Planets Assessment Group (OPAG) of the Lunar and Planetary Institute*. Universities Space Research Association. p. 33.

[7] Barnett, Amanda (28 August 2015). "Pluto probe gets new assignment". *CNN*. Retrieved 7 October 2015.

[8] Dunn, Marcia (22 October 2015). "NASA's New Horizons on new post-Pluto mission". *AP News*. Retrieved 25 October 2015.

[9] "NASA's New Horizons Completes Record-Setting Kuiper Belt Targeting Maneuvers". New Horizons Team. 5 November 2015. Retrieved 6 November 2015.

[10] Lakdawalla, Emily (1 September 2015). "New Horizons extended mission target selected". *Planetary Society blog*. Planetary Society.

[11] J. R. Spencer, M. W. Buie; et al. (2015). "The Successful Search for a Post-Pluto KBO Flyby Target for New Horizons Using the Hubble Space Telescope" (PDF). *European Planetary Science Congress (EPSC) Abstract* (Copernicus Office).

[12] "Hubble Survey Finds Two Kuiper Belt Objects to Support New Horizons Mission". *HubbleSite news release*. Space Telescope Science Institute. 1 July 2014.

[13] "NASA's Hubble Telescope Finds Potential Kuiper Belt Targets for New Horizons Pluto Mission". *HubbleSite*. 15 October 2014.

[14] Wall, Mike (15 October 2014). "Hubble Telescope Spots Post-Pluto Targets for New Horizons Probe". Space.com. Archived from the original on 15 October 2014.

[15] Porter, S. B.; Parker, A. H.; et al. (eds.). *Orbits and Accessibility of Potential New Horizons KBO Encounter Targets* (PDF). 46th Lunar and Planetary Science Conference (2015).

[16] "Maneuver Moves New Horizons Spacecraft toward Next Potential Target". 23 October 2015. Retrieved 5 November 2015.

[17] "New Horizons Continues Toward Potential Kuiper Belt Target". 26 October 2015. Retrieved 5 November 2015.

[18] "On Track: New Horizons Carries Out Third KBO Targeting Maneuver". 29 October 2015. Retrieved 5 November 2015.

[19]

16.7 External links

- 2014 MU69 at the *JPL Small-Body Database*

  - Discovery · Orbit diagram · Orbital elements · Physical parameters

Chapter 17

2014 OS393

2014 OS$_{393}$ (formerly labeled **e31007AI** in the context of the Hubble Space Telescope, and **e3** and **PT2** in the context of the *New Horizons* mission) is a Kuiper belt object (KBO) and formerly a potential flyby target for the *New Horizons* probe.[3]

17.1 Discovery and naming

2014 OS$_{393}$ was discovered with the help of the Hubble Space Telescope (HST)[6] because the object has a magnitude of 26.3, which is too faint to be observed by ground-based telescopes. Preliminary observations by the HST searching for KBO flyby targets for the *New Horizons* probe started in June 2014, and more intensive observations continued in July and August.[7][8] 2014 OS$_{393}$ was first discovered in observations on July 30, 2014, but it was designated e31007AI at the time, nicknamed e3 for short.[3][5] Its existence as a potential target of the *New Horizons* probe was revealed by NASA in October 2014[9][10] and designated PT2, but the official name 2014 OS$_{393}$ was not assigned by the Minor Planet Center (MPC) until March 2015 after better orbit information was available.[3]

17.2 Potential targets of the *New Horizons* mission

After the *New Horizons* probe completed its flyby of Pluto, the probe is to be manoeuvred to a flyby of at least one Kuiper belt object (KBO). Several potential targets were under consideration for the first such flyby. Potential target PT2, the KBO 2014 OS$_{393}$, has a diameter between 30–55 km (19–34 mi) and the potential encounter in 2018–2019 would have been at a distance of 43–44 AU from the Sun.[2] The potential targets for the *New Horizons* probe are PT1 and PT3, the KBOs 2014 MU69 and 2014 PN70, and the probe has sufficient fuel to maneuver to either PT1 or PT3. Potential target PT2 is no longer under consideration as a potential target,[11] and 2014 MT69 was eliminated as a target before the fall of 2014.[3]

On 28 August 2015, the *New Horizons* team announced the selection of 2014 MU$_{69}$ as the next flyby target.[12]

17.3 References

[1] "2014 OS393". Minor Planet Center. 2014-10-24.

[2] Lakdawalla, Emily (October 15, 2014). "Finally! New Horizons has a second target". *Planetary Society blog*. Planetary Society. Archived from the original on October 15, 2014.

[3] Zangari, Amanda (March 28, 2015). "Postcards from Pluto". Tumblr.

[4] "JPL Small-Body Database Browser: (2014 OS393)" (2014-10-24 last obs; arc: 86 days). Retrieved 8 May 2015.

[5] Buie, Marc (October 15, 2014). "New Horizons HST KBO Search Results: Status Report" (PDF). Space Telescope Science Institute. p. 23.

[6] J. R. Spencer, M. W. Buie; et al. (2015). "The Successful Search for a Post-Pluto KBO Flyby Target for New Horizons Using the Hubble Space Telescope" (PDF). *European Planetary Science Congress (EPSC) Abstract* (Copernicus Office).

[7] "Hubble to Proceed with Full Search for New Horizons Targets". *HubbleSite news release.* Space Telescope Science Institute. July 1, 2014.

[8] Schmidt, Klaus (2 July 2014). "*Hubble* to Proceed with Full Search for *New Horizons* Targets". *International Space Fellowship.*

[9] "NASA's Hubble Telescope Finds Potential Kuiper Belt Targets for New Horizons Pluto Mission". *HubbleSite.* 15 October 2014.

[10] Wall, Mike (October 15, 2014). "Hubble Telescope Spots Post-Pluto Targets for New Horizons Probe". Space.com. Archived from the original on October 15, 2014.

[11] Powell, Corey S. (March 29, 2015). "Alan Stern on Pluto's Wonders, New Horizons' Lost Twin, and That Whole "Dwarf Planet" Thing". *Discover.*

[12] Cofield, Calla (28 August 2015). "Beyond Pluto: 2nd Target Chosen for New Horizons Probe". *Space.com.*

Chapter 18

2014 PN70

2014 PN$_{70}$ (formerly labeled **g12000JZ** in the context of the Hubble Space Telescope, and **g1** and **PT3** in the context of the *New Horizons* mission) is a Kuiper belt object (KBO) and was a proposed flyby target for the *New Horizons* probe.[3]

18.1 Discovery and naming

2014 PN$_{70}$ was discovered during an observation campaign intended to search for KBO flyby targets for the *New Horizons* probe.[8] The observations started in June 2014, and more intensive ones continued in July and August.[9] They were conducted with the help of the Hubble Space Telescope (HST); 2014 PN$_{70}$'s magnitude of 26.4 is too faint to be observed by ground-based telescopes. 2014 PN$_{70}$ was first discovered in observations on August 6, 2014, and it was designated g12000JZ at the time, nicknamed g1 for short.[3][7] Its existence as a potential target of the *New Horizons* probe was revealed by NASA in October 2014[10][11] and it was designated PT3; its official name, 2014 PN$_{70}$, was not assigned by the Minor Planet Center (MPC) until March 2015 after better orbit information was available.[3]

18.2 Characteristics

2014 PN$_{70}$, has a diameter between 35–120 km (22–75 mi).[2]

18.3 Exploration

Having completed its flyby of Pluto, the *New Horizons* space probe will be maneuvered to a flyby of at least one Kuiper belt object (KBO). Several potential targets were under consideration. 2014 PN$_{70}$ (PT3) was considered a second choice after 2014 MU69 (PT1), because more fuel was required to carry out a flyby. 2014 OS393 (PT2) was already no longer under consideration as a potential target.[12]

On 28 August 2015, the *New Horizons* team announced the selection of 2014 MU$_{69}$ as the next flyby target.[13]

18.4 References

[1] "2014 PN70". Minor Planet Center. 2014-10-22.

[2] Lakdawalla, Emily (October 15, 2014). "Finally! New Horizons has a second target". *Planetary Society blog*. Planetary Society. Archived from the original on October 15, 2014.

[3] Zangari, Amanda (March 28, 2015). "Postcards from Pluto". Tumblr.

[4] "JPL Small-Body Database Browser: (2014 PN70)" (2014-10-22 last obs; arc: 77 days). Retrieved 7 May 2015.

[5] Stern, Alan (August 2015). "OPAG: We Did It!" (PDF). *Presentation to the Outer Planets Assessment Group (OPAG) of the Lunar and Planetary Institute.* Universities Space Research Association. p. 33.

[6] "ABSOLUTE MAGNITUDE (H)". NASA.

[7] Buie, Marc (October 15, 2014). "New Horizons HST KBO Search Results: Status Report" (PDF). Space Telescope Science Institute. p. 23.

[8] J. R. Spencer, M. W. Buie; et al. (2015). "The Successful Search for a Post-Pluto KBO Flyby Target for New Horizons Using the Hubble Space Telescope" (PDF). *European Planetary Science Congress (EPSC) Abstract* (Copernicus Office).

[9] "Hubble to Proceed with Full Search for New Horizons Targets". *HubbleSite news release.* Space Telescope Science Institute. July 1, 2014.

[10] "NASA's Hubble Telescope Finds Potential Kuiper Belt Targets for New Horizons Pluto Mission". *HubbleSite.* 15 October 2014.

[11] Wall, Mike (October 15, 2014). "Hubble Telescope Spots Post-Pluto Targets for New Horizons Probe". Space.com. Archived from the original on October 15, 2014.

[12] Powell, Corey S. (March 29, 2015). "Alan Stern on Pluto's Wonders, New Horizons' Lost Twin, and That Whole "Dwarf Planet" Thing". *Discover.*

[13] Cofield, Calla (28 August 2015). "Beyond Pluto: 2nd Target Chosen for New Horizons Probe". *Space.com.*

Chapter 19

2014 UM33

2014 UM$_{33}$ (2010 TQ$_{182}$) is a possible dwarf planet and trans-Neptunian object residing in the outer Kuiper belt. It was discovered on October 22, 2014 by the Mount Lemmon Survey. Its orbit was initially poorly determined, with 17 observations over 62 days, giving it an orbital uncertainty of 8. It is listed on Mike Brown's website as a probable dwarf planet, ranked 56th most likely.[4] It is approximately the size of 2 Pallas in the asteroid belt. On August 18, 2015, 2014 UM33 was found to have been discovered over four years previously, with the designation 2010 TQ$_{182}$. This extended its observation arc to over 4 years, and then precovery observations were found using the Sloan Digital Sky Survey from 2009.

19.1 References

[1] "IAU Minor Planet Center - 2014 UM33". *Minor Planet Center*. International Astronomical Union. Retrieved 2 January 2015.

[2] "Jet Propulsion Laboratory - Small Body Database". *Jet Propulsion Laboratory*. NASA. Retrieved 2 January 2015.

[3] "Glossary: Absolute Magnitude (H)". *JPL*. NASA. Retrieved 2 January 2015.

[4] Brown, Mike. "How many dwarf planets are there in the outer solar system? (updates daily)". *http://web.gps.caltech.edu/ ~{}mbrown*. Caltech. Retrieved 21 July 2015.

[5] Wm. Robert Johnston (24 March 2015). "List of Known Trans-Neptunian Objects". Johnston's Archive. Retrieved 21 July 2015.

[6] Bruton, Dan. "Conversion of Absolute Magnitude to Diameter". *http://www.physics.sfasu.edu*. sfasu.edu. Retrieved 18 August 2015.

Chapter 20

2011 KW48

2011 KW$_{48}$ (also labeled VNH0004) is a Kuiper belt object. In January 2015, the New Horizons probe traveled about 75 million km (47 million miles) (0.5 AU) from it,[2] which is too far to take high-resolution photos of the object itself, but allowed the detection of possible satellites. If 2011 KW$_{48}$ were 100 kilometers wide, it would appear approximately 0.11 arcseconds wide to *New Horizons*.

2011 KW$_{48}$ was only observed 12 times by the Mauna Kea (8) and Las Campanas Observatory (4) over a period of about 33.8 days between May 29 and July 2, 2011, so its current orbit is extremely uncertain. Between January 4th and 15th in 2015 *New Horizons* actively observed the object.[3]

20.1 References

[1] "Orbit Fit and Astrometric record for VNH0004". *http://www.boulder.swri.edu*. Retrieved October 25, 2014.

[2] "New Horizons to Encounter KBO Ahead of Pluto Flyby". *AmericaSpace*.

[3] Gebhardt, Chris (January 19, 2015). "New Horizons begins Pluto observations ahead of July flyby". *Nasaspaceflight.com*. Retrieved June 7, 2015.

Chapter 21

(230965) 2004 XA192

(230965) 2004 XA$_{192}$ is a Kuiper-belt object with a diameter of 339+120
−95 km. It has an absolute magnitude of 4.11. It was discovered on 12 December 2004 at Palomar Observatory.

It is currently at 35.8 AU from the Sun, near its perihelion.[5]

21.1 References

[1] Marc Buie (2014-08-23). "Orbit Fit and Astrometric record for 230965". SwRI (Space Science Department). Retrieved 2014-11-14.

[2] "JPL Small-Body Database Browser: 230965 (2004 XA192)" (2011-11-22 last obs; arc: 22.23 years). Retrieved 2014-11-14.

[3] TNOs are Cool: A survey of the trans-Neptunian region. X. Analysis of classical Kuiper belt objects from Herschel* and Spitzer observations p. 18

[4] 0 & n = 230965 AstDys Summary for (230965) 2004 XA192 Retrieved 31 August 2009

[5] AstDys Ephmerides for (230965) 2004 XA192, Retrieved 31 August 2009

21.2 External links

- Chart Trajectory by JPL (software needs JAVA)

Chapter 22

(420356) 2012 BX85

(420356) 2012 BX$_{85}$ is a trans-Neptunian object (TNO) and possibly a dwarf planet[2] lying in the Kuiper Belt. It has the second lowest eccentricity of any TNO, after 2003 YN$_{179}$. (420356) 2012 BX$_{85}$ orbits near the 3:5 resonance to Neptune, but takes about 160 Neptune orbits (26,500 years) to make an orbit less than would be expected of an object in a true 3:5 resonance. It was discovered on January 23, 2012, with precovery observations accepted by the Minor Planet Center dating back to December 2011, with possible precovery observations dating back to 2004 that have yet to be accepted.

The asteroid is one of the most recently-discovered asteroids to be given a numeric designation, most likely because of the large number of observations, on average about one every 10 days since it was discovered. All of the observations, except 6 in February 2013, were made by the Mount Graham Observatory, the discovery site.

22.1   References

[1] "2012 BX85 - JPL Small Body Database". *JPL* (2015-01-17 last obs). NASA. Retrieved September 2015.

[2] Brown, Mike. "How many dwarf planets are there in the Solar System". Retrieved 2 February 2015.

Chapter 23

Actaea (moon)

Actaea, officially **(120347) Salacia I Actaea**, is the single known natural satellite of the cubewano 120347 Salacia. Its diameter is estimated 300 km, which is ⅓ of the diameter of Salacia. Therefore, it is seen as binary asteroid. Actaea is about the sixth-biggest known moon of a Kuiper belt object, after Charon (1212 km), Dysnomia (685 km), Vanth (378 km), Ilmarë (361 km) and Hiiaka (320 km).

23.1 Discovery and name

It was discovered on 21 July 2006 by Keith S. Noll, Harold Levison, Denise Stephens and Will Grundy with the Hubble Space Telescope.[1] On 18 February 2011, it was officially named Actaea after the nereid Aktaia.

23.2 Orbit

Actaea orbits its primary every 5.49380±0.00016 d at a distance of 5619±87 km and with an eccentricity of 0.0084±0.0076.[2] The ratio of its semi-major axis to its primary's Hill radius is 0.0023, the tightest trans-Neptunian binary with a known orbit.[3]

23.3 Physical characteristics

Actaea is 2.372±0.060 magnitudes fainter than Salacia,[4] implying a diameter ratio of 2.98 for equal albedos.[3] Hence, assuming equal albedos, it has a diameter of 303±35 km[4] Actaea has the same color as Salacia (V–I = 0.89±0.02 and 0.87±0.01, respectively), supporting the assumption of equal albedos.[3] It has been calculated that the Salacia system should have undergone enough tidal evolution to circularize their orbits, which is consistent with the low measured eccentricity, but that the primary need not have been tidally locked.[3] According to the low density (1,16 g/cm^3), it is expected that Actaea consists of water ice. Salacia and Actaea will next occult each other in 2067.[3] The mass of the system is 4,66 ± 0,22 × 10^{20} kg, 4% amounts the mass of Actaea.[3]

23.4 References

[1] "IAUC 8751: (120347) 2004 SB_60; 2006gi, 2006gj; V733 Cep". Cbat.eps.harvard.edu. Retrieved 2014-06-14.

[2] Johnston Archive: (120347) Salacia and Actaea

[3] Stansberry, J.A.; Grundy, W.M.; Mueller, M.; et al. (2012). "Physical Properties of Trans-Neptunian Binaries (120347) Salacia–Actaea and (42355) Typhon–Echidna". *Icarus* **219**: 676–688. Bibcode:2012Icar..219..676S. doi:10.1016/j.icarus.2012.03.029. CiteSeerX: 10.1.1.398.6675.

[4] Fornasier, S.; Lellouch, E.; Müller, P., T.; *et .al.* (2013). "TNOs are Cool: A survey of the trans-Neptunian region. VIII. Combined Herschel PACS and SPIRE observations of 9 bright targets at 70–500 μm". *Astronomy&Astrophysics* **555**: A92. arXiv:1305.0449v2. Bibcode:2013A&A...555A..15F. doi:10.1051/0004-6361/201321329.

Chapter 24

Heliosphere

This article is about the Sun's astrosphere. For astrospheres of other stars, see stellar-wind bubble.

The **heliosphere** is the bubble-like region of space dominated by the Sun, which extends far beyond the orbit of Pluto. Plasma "blown" out from the Sun, known as the solar wind, creates and maintains this bubble against the outside pressure of the interstellar medium, the hydrogen and helium gas that permeates the Milky Way Galaxy. The solar wind flows outward from the Sun until encountering the **termination shock**, where motion slows abruptly. The Voyager spacecraft have actively explored the outer reaches of the heliosphere, passing through the shock and entering the **heliosheath**, a transitional region which is in turn bounded by the outermost edge of the heliosphere, called the **heliopause**. The overall shape of the heliosphere is controlled by the interstellar medium, through which it is traveling, as well as the Sun, and does not appear to be perfectly spherical.[1] The limited data available and unexplored nature[2] of these structures have resulted in many theories.

On September 12, 2013, NASA announced that Voyager 1 had exited the heliosphere on August 25, 2012, when it measured a sudden increase in plasma density of about forty times.[3] Because the heliopause marks one boundary between the Sun's solar wind and the rest of the galaxy, a spacecraft such as Voyager 1 which has departed the heliosphere can be said to have reached interstellar space.

24.1 Summary

Except for localized regions near obstacles such as planets or comets, the heliosphere is dominated by material emanating from the Sun, although cosmic rays and fast-moving neutral atoms can penetrate the heliosphere from the outside. Originating at the extremely hot surface of the corona, solar wind particles reach escape velocity, streaming outwards at 300 to 800 km/s (671 thousand to 1.79 million mph or 1 to 2.9 million km/h).[4] As it begins to interact with the interstellar medium, its velocity slows before finally stopping altogether. The point where the solar wind becomes slower than the speed of sound is called the termination shock; the solar wind continues to slow as it passes through the heliosheath leading to a boundary called the heliopause, where the interstellar medium and solar wind pressures balance. The termination shock was traversed by Voyager 1 in 2004,[5] and Voyager 2 in 2007.[1]

It was thought that beyond the heliopause there was a bow shock, but data from Interstellar Boundary Explorer suggested the velocity of the Sun through the interstellar medium is too low for it to form.[6] It may be a more gentle "bow wave".[7] Voyager data led to a new theory that the heliosheath has "magnetic bubbles" and a stagnation zone.[8][9]

The 'stagnation region' within the heliosheath, starting around 113 AU, was detected by Voyager 1 in 2010.[8] There the solar wind velocity drops to zero, the magnetic field intensity doubles, and high-energy electrons from the galaxy increase 100-fold.[8] Starting in May 2012 at 120 AU, Voyager 1 detected a sudden increase in cosmic rays, an apparent signature of approach to the heliopause.[10] In December 2012 NASA announced that in late August 2012 Voyager 1, at about 122 AU from the Sun, entered a new region they called the "magnetic highway", an area still under the influence of the Sun, but with some dramatic differences.[5] In the summer of 2013 NASA announced that Voyager 1 had reached interstellar

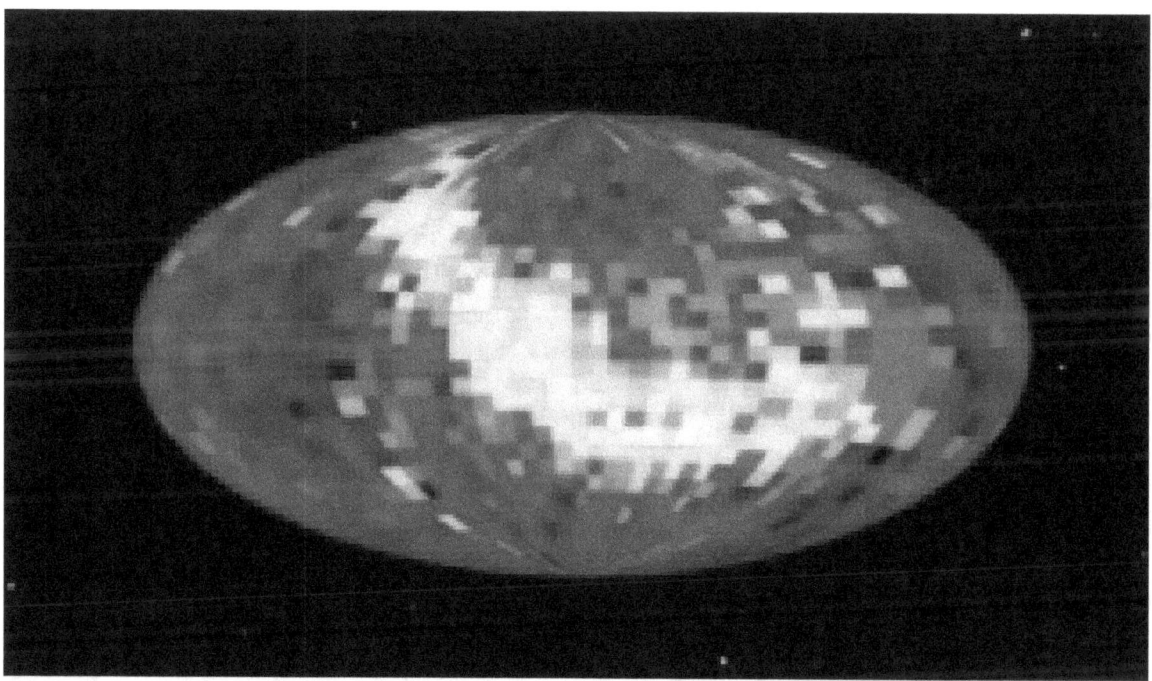

Energetic neutral atoms map by IBEX. Credit: NASA / Goddard Space Flight Center Scientific Visualization Studio.

space as of August 25, 2012.[11]

Cassini and IBEX data challenged the "heliotail" theory in 2009.[12][13] In July 2013, IBEX results revealed a 4-lobed tail on the Solar System's heliosphere.[14]

24.2 Solar wind

Main articles: Solar wind and Interplanetary medium

The solar wind consists of particles (ionized atoms from the solar corona) and fields (in particular, magnetic fields). As the Sun rotates once in approximately 25 days, the magnetic field transported by the solar wind gets wrapped into a spiral. Variations in the Sun's magnetic field are carried outward by the solar wind and can produce magnetic storms in the Earth's own magnetosphere.

24.3 Structure

24.3.1 Heliospheric current sheet

Main article: Heliospheric current sheet

The heliospheric current sheet is a ripple in the heliosphere created by the rotating magnetic field of the Sun. Extending throughout the heliosphere, it is considered the largest structure in the Solar System and is said to resemble a "ballerina's skirt".[15]

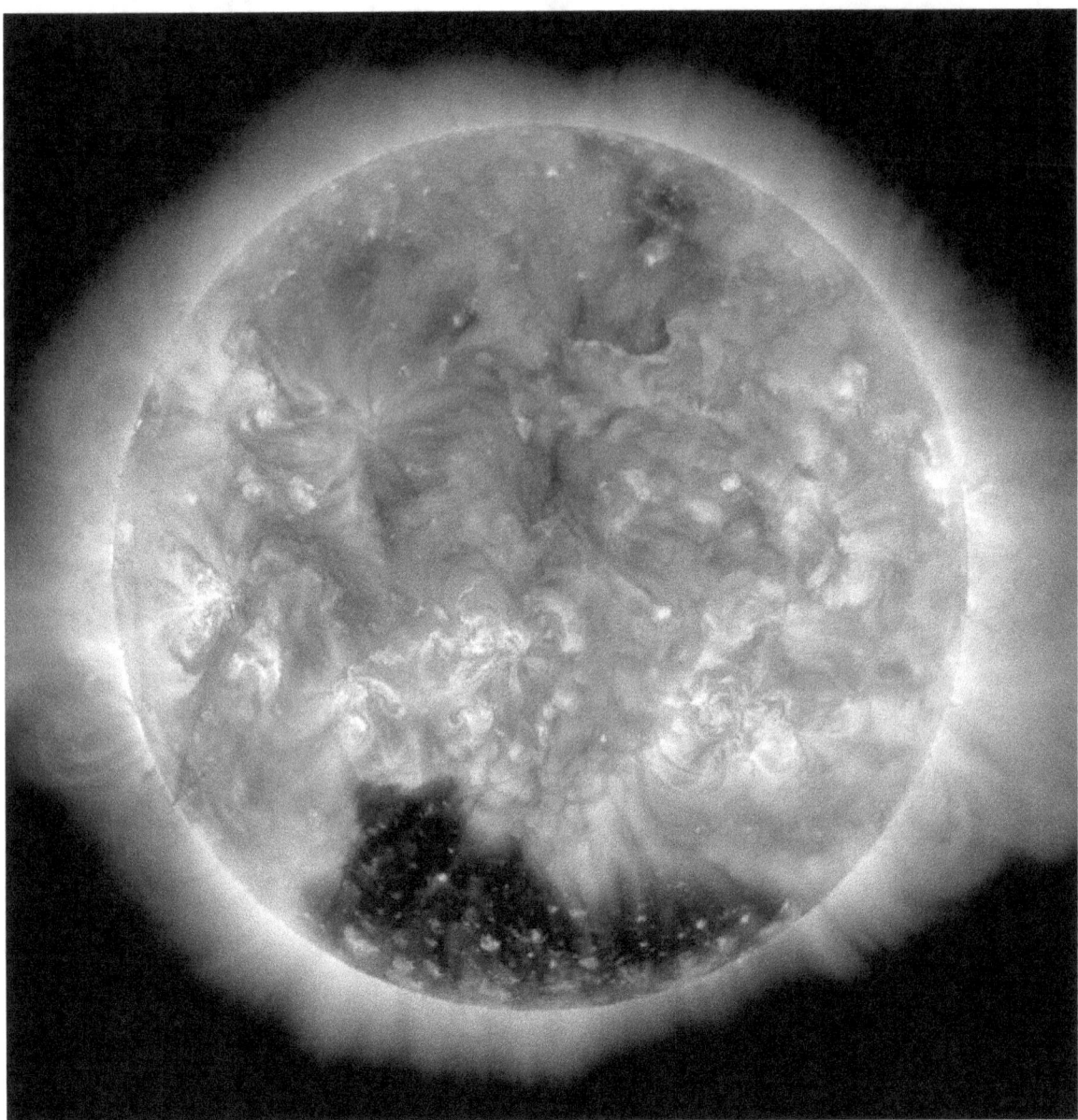

Earth's Sun as seen at a wavelength of 193 Ångströms

24.3.2   Outer structure

The outer structure of the heliosphere is determined by the interactions between the solar wind and the winds of interstellar space. The solar wind streams away from the Sun in all directions at speeds of several hundred km/s in the Earth's vicinity. At some distance from the Sun, well beyond the orbit of Neptune, this supersonic wind must slow down to meet the gases in the interstellar medium. This takes place in several stages:

- The solar wind is traveling at supersonic speeds within the Solar System. At the termination shock, a standing shock wave, the solar wind falls below the speed of sound and becomes subsonic.

- It was previously held that, once subsonic, the solar wind might be affected by the ambient flow of the interstellar medium: Its pressures were theorized to cause the solar wind to form a nose on one side and comet-like heliotail behind. The area called the heliosheath. However, scientific results in 2009 showed that this model is incorrect.[12][13] As of 2011, it is thought to be filled with a magnetic bubble "foam".[16]

The heliospheric current sheet out to the orbit of Jupiter

- The outer surface of the heliosheath, where the heliosphere meets the interstellar medium, is called the **heliopause**. This is the edge of the entire heliosphere. Scientific results in 2009 adjusted this model.[12][13]

- In theory, the heliopause causes turbulence in the interstellar medium as the sun orbits the Galactic Center. Outside the heliopause, would be a turbulent region caused by the pressure of the advancing heliopause against the interstellar medium. However, the velocity of Solar wind relative to the interstellar medium is probably too low for a bow shock.[6]

24.3.3 Termination shock

The termination shock is the point in the heliosphere where the solar wind slows down to subsonic speed (relative to the Sun) because of interactions with the local interstellar medium. This causes compression, heating, and a change in the magnetic field. In the Solar System the termination shock is believed to be 75 to 90 astronomical units[17] from the Sun. In 2004, Voyager 1 crossed the Sun's termination shock followed by Voyager 2 in 2007.[18][19][20] [21][22][3] [23][1]

The shock arises because solar wind particles are emitted from the Sun at about 400 km/s, while the speed of sound (in the interstellar medium) is about 100 km/s. (The exact speed depends on the density, which fluctuates considerably.) The interstellar medium, although very low in density, nonetheless has a constant pressure associated with it; the pressure from the solar wind decreases with the square of the distance from the Sun. As one moves far enough away from the Sun, the pressure from the interstellar medium becomes equal to the pressure from solar wind, at which point the solar wind slows to below its speed of sound, causing a shock wave.

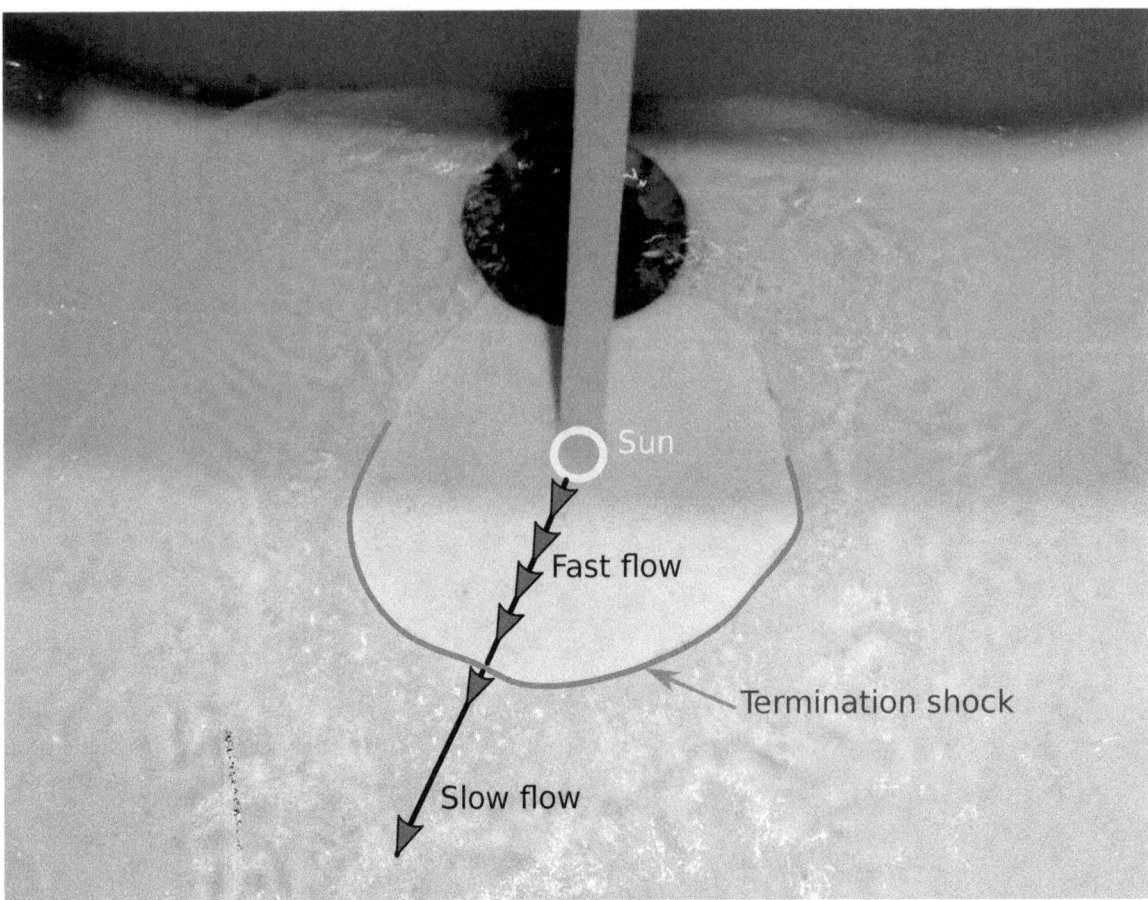

A termination shock in a sink basin

Other termination shocks can be seen in terrestrial systems; perhaps the easiest may be seen by simply running a water tap into a sink creating a hydraulic jump. Upon hitting the floor of the sink, the flowing water spreads out at a speed that is higher than the local wave speed, forming a disk of shallow, rapidly diverging flow (analogous to the tenuous, supersonic solar wind). Around the periphery of the disk, a shock front or wall of water forms; outside the shock front, the water moves slower than the local wave speed (analogous to the subsonic interstellar medium).

Going outward from the Sun, the termination shock is followed by the heliopause, where solar wind particles are stopped by the interstellar medium.

Evidence presented at a meeting of the American Geophysical Union in May 2005 by Ed Stone suggests that the *Voyager 1* spacecraft passed the termination shock in December 2004, when it was about 94 AU from the Sun, by virtue of the change in magnetic readings taken from the craft. In contrast, *Voyager 2* began detecting returning particles when it was only 76 AU from the Sun, in May 2006. This implies that the heliosphere may be irregularly shaped, bulging outwards in the Sun's northern hemisphere and pushed inward in the south.[24]

24.3.4 Heliosheath

The heliosheath is the region of the heliosphere beyond the termination shock. Here the wind is slowed, compressed and made turbulent by its interaction with the interstellar medium. Its distance from the Sun is approximately 80 to 100 astronomical units (AU) at its closest point.

A proposed model hypothesizes that the heliosheath is shaped like the coma of a comet, and trails several times that distance in the direction opposite to the Sun's path through space. At its windward side, its thickness is estimated to be between 10 and 100 AU.[26] However, scientific results in 2009 showed that model may be incorrect.[12][13]

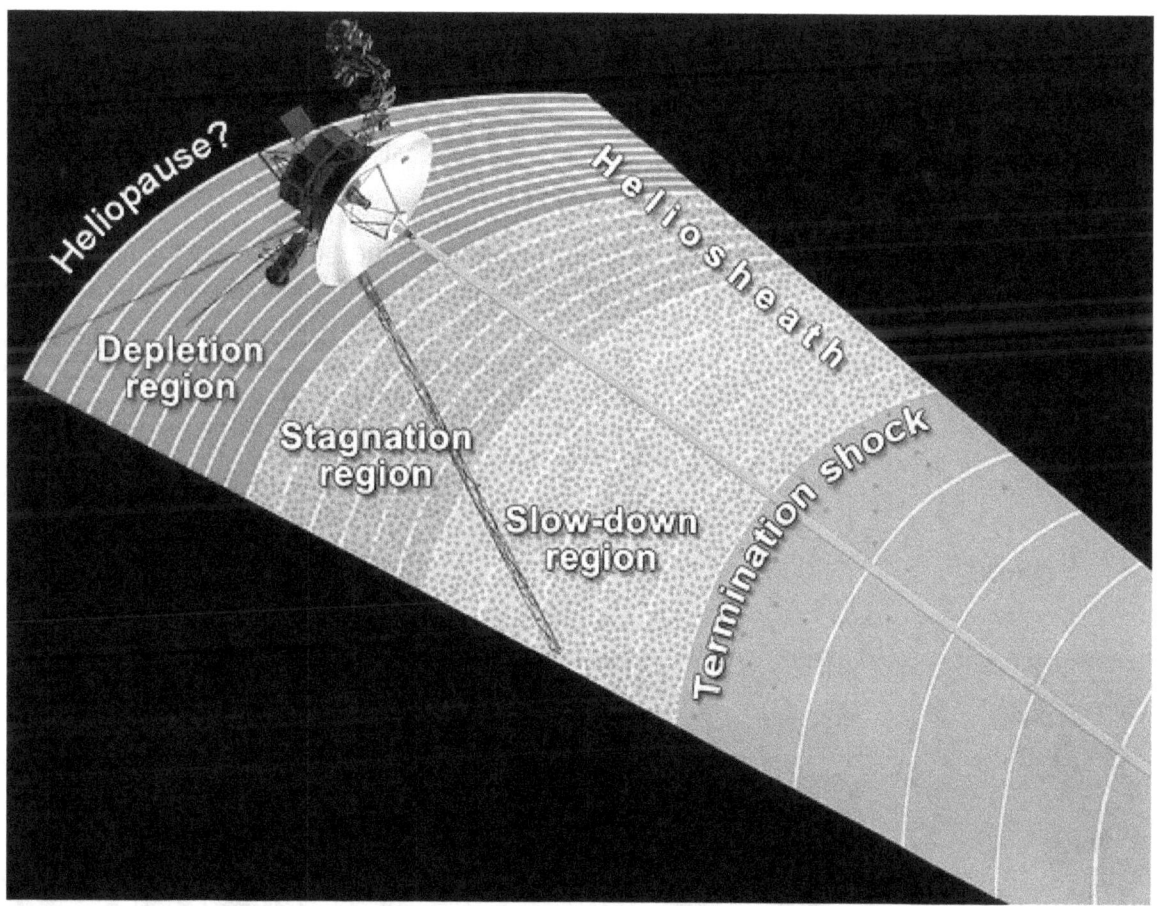

This picture about the heliosphere was released on June 28, 2013 and incorporates results from the Voyager spacecraft.[25]

The *Voyager 1* and *Voyager 2* spacecraft are currently studying the heliosheath. In late 2010, *Voyager 1* reached a region of the heliosheath where the solar wind's velocity had dropped to zero.[27][28][29][30] In 2011, astronomers announced that the *Voyagers* had determined that the heliosheath is not smooth, but is filled with 100 million-mile-wide bubbles created by the impact of the solar wind and the interstellar medium.[31][32] *Voyager 1* and *2* began detecting evidence for the bubbles in 2007 and 2008, respectively.[32] The probably sausage-shaped bubbles are formed by magnetic reconnection between oppositely oriented sectors of the solar magnetic field as the solar wind slows down.[32] They probably represent self-contained structures that have detached from the interplanetary magnetic field.[31][32]

24.3.5 Heliopause

The heliopause is the theoretical boundary where the Sun's solar wind is stopped by the interstellar medium; where the solar wind's strength is no longer great enough to push back the stellar winds of the surrounding stars. This is the boundary where the interstellar medium and solar wind pressures balance. The crossing of the heliopause should be signaled by a sharp drop in the temperature of charged particles,[28] a change in the direction of the magnetic field, and an increase in the amount of galactic cosmic rays.[10] In May 2012, *Voyager 1* detected a rapid increase in such cosmic rays (a 9% increase in a month, following a more gradual increase of 25% from Jan. 2009 to Jan. 2012), suggesting it was approaching the heliopause.[10] In the fall of 2013, NASA announced that Voyager 1 had crossed the heliopause as of August 25, 2012.[11] This was at a distance of 121 AU (18 billion km) from the Sun.[33] Contrary to predictions, data from Voyager 1 indicates the magnetic field of the galaxy is aligned with the solar magnetic field.[34]

24.3.6 Heliotail

The heliotail is the solar system's tail or can be understood as the tail of the heliosphere. Similarly it can be compared to a comet, which also has a tail (however a comet's tail does not stretch behind it as it moves, it is always pointing away from the Sun). A further explanation of the tail is a region where the Sun's million mile per hour solar wind flows down and ultimately escapes the heliosphere, slowly evaporating because of charge exchange.[35] The shape of this newly found tail by NASA's Interstellar Boundary Explorer (IBEX) is that of a four-leaf clover.[36] Due to the particles in the tail, they do not shine, therefore it cannot be seen with conventional instruments. IBEX has made the first observations by using a technique called "energetic neutral atom energy" which is the process of measuring the neutral particles created by collisions at the solar system's boundaries.[36]

The tail has been shown to contain fast and slow particles; the slow particles are on the side and the fast particles are encompassed in the center. The shape of the tail can be linked to the sun sending out fast solar winds near its poles and slow solar wind near its equator more recently. The clover-shaped tail moves further away from the sun, which makes the charged particles begin to morph into a new orientation.

24.4   Beyond

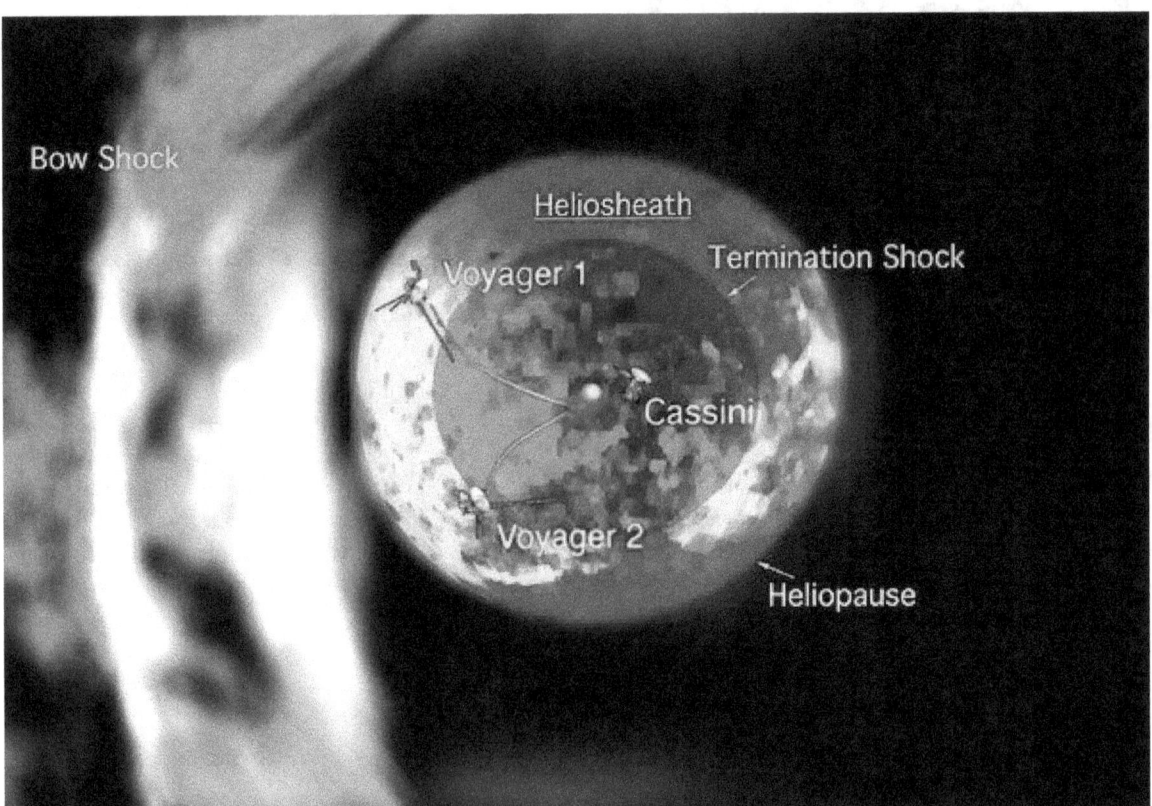

The bubble-like heliosphere moving through the interstellar medium

The heliopause is the final known boundary between the heliosphere and the interstellar space that is filled with material, especially plasma, not from our own star, the Sun, but from other stars.[37] Even so, just outside the heliosphere (i.e. the "solar bubble") there is a transitional region, as detected by Voyager 1.[38] Just as some interstellar pressure was detected as early as 2004, some of the Sun's material seeps into the interstellar medium.[38] The heliosphere is thought to reside in the Local Interstellar Cloud inside the Local Bubble, which is a region in the Orion Arm of the Milky Way Galaxy.

Outside the heliosphere there is a forty-fold increase in plasma density.[38] There is also a radical reduction in the detection

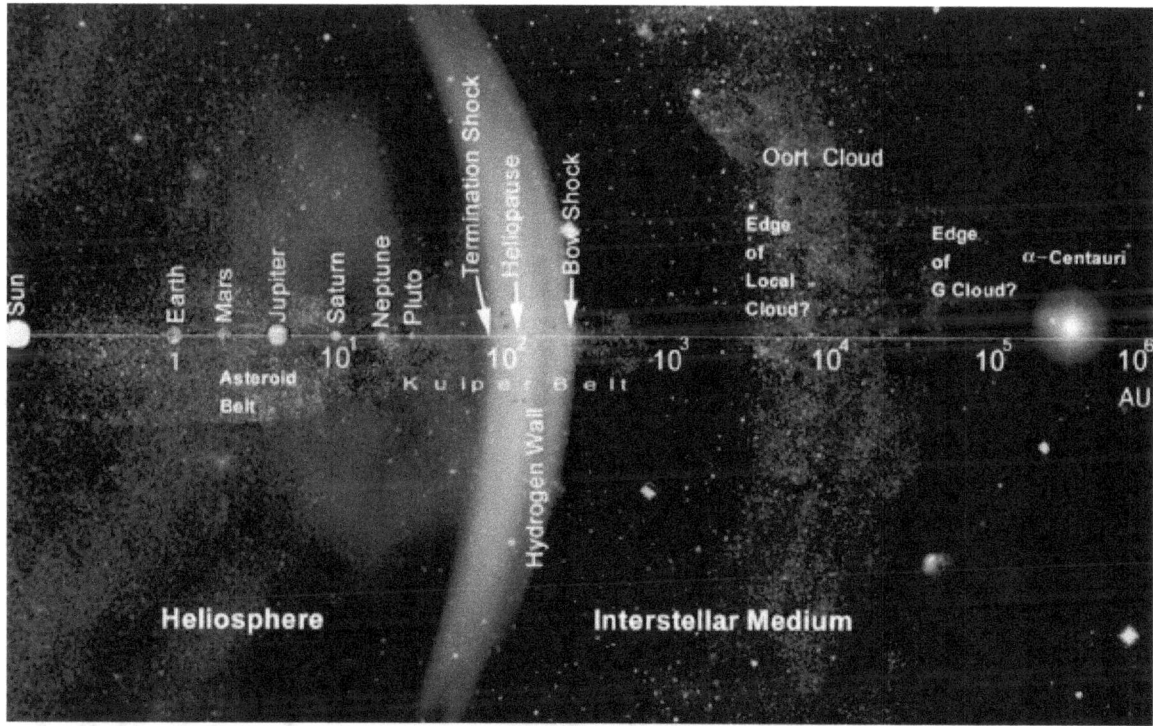

Solar map with the location of the hypothetical hydrogen wall and bow shock

of certain types of particles from the Sun, and a large increase in Galactic cosmic rays.[39]

The flow of the interstellar medium (ISM) into the heliosphere has been measured by at least 11 different spacecraft as of 2013.[40] By 2013, it was suspected that the direction of the flow had changed over time.[40] The flow, coming from Earth's perspective from the constellation Scorpius, has probably changed direction by several degrees since the 1970s.[40]

24.4.1 Hydrogen wall

"Hydrogen wall" redirects here. For other topics, see Hydrogen (disambiguation).

According to one hypothesis,[41] there exists a region of hot hydrogen known as the **hydrogen wall** between the bow shock and the heliopause. The wall is composed of interstellar material interacting with the edge of the heliosphere. One paper released in 2013 studied the concept of a bow wave and hydrogen wall.[7]

Another hypothesis suggests that the heliopause could be smaller on the side of the Solar System facing the Sun's orbital motion through the galaxy. It may also vary depending on the current velocity of the solar wind and the local density of the interstellar medium. It is known to lie far outside the orbit of Neptune. The current mission of the *Voyager 1* and *2* spacecraft is to find and study the termination shock, heliosheath, and heliopause. Meanwhile, the Interstellar Boundary Explorer (IBEX) mission is attempting to image the heliopause from Earth orbit within two years of its 2008 launch. Initial results (October 2009) from IBEX suggest that previous assumptions are insufficiently cognisant of the true complexities of the heliopause.[42]

When particles emitted by the sun bump into the interstellar ones, they slow down while releasing energy. Many particles accumulate in and around the heliopause, highly energised by their negative acceleration, creating a shock wave. An alternative definition is that the heliopause is the magnetopause between the Solar System's magnetosphere and the galaxy's plasma currents.

24.4.2 Bow shock

Further information: Bow shocks in astrophysics

It was long hypothesized that the Sun produces a "shock wave" in its travels within the ISM. It would occur if the interstellar medium is moving supersonically "toward" the Sun, since its solar wind moves "away" from the Sun supersonically. When the interstellar wind hits the heliosphere it slows and creates a region of turbulence. A bow shock was thought to possibly occur at about 230 AU,[17] but in 2012 it was determined it probably does not exist.[6] This conclusion resulted from new measurements: The velocity of the LISM (Local Interstellar Medium) relative to the Sun's was previously measured to be 26.3 km/s by Ulysses, whereas IBEX measured it at 23.2 km/s.[43]

This phenomenon has been observed outside our solar system, around stars other than the Sun, by NASA's now retired orbital GALEX telescope. The red giant star Mira in the constellation Cetus has been shown to have both a debris tail of ejecta from the star and a distinct shock in the direction of its movement through space (at over 130 kilometers per second).

24.5   Observational methods

Pioneer H, shown here in a Museum, was a canceled probe to study the Sun[44]

24.5.1 Detection by spacecraft

The precise distance to, and shape of the heliopause is still uncertain. Interplanetary/interstellar spacecraft such as *Pioneer 10*, *Pioneer 11* and *Voyager 2* are traveling outward through the Solar System and will eventually pass through the heliopause.

Cassini results

Rather than a comet-like shape, the heliosphere appears to be bubble-shaped according to data from Cassini's Ion and Neutral Camera (MIMI / INCA). Rather than being dominated by the collisions between the solar wind and the interstellar medium, the INCA (ENA) maps suggest that the interaction is controlled more by particle pressure and magnetic field energy density.[12] [45] The new shape from the data is thought more like a spherical bubble, than a cometary shape.[12]

IBEX results

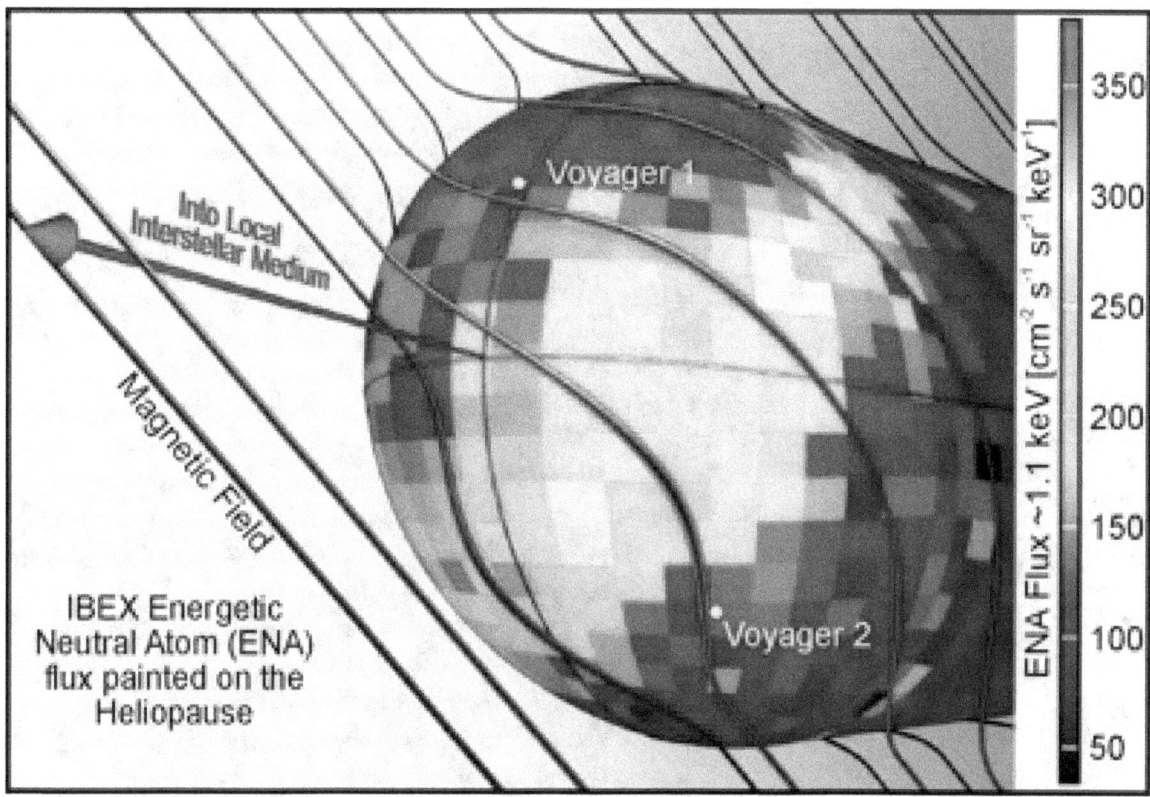

Initial data from Interstellar Boundary Explorer (IBEX), launched in October 2008, revealed a previously unpredicted "very narrow ribbon that is two to three times brighter than anything else in the sky."[13] Initial interpretations suggest that "the interstellar environment has far more influence on structuring the heliosphere than anyone previously believed"[46] "No one knows what is creating the ENA (energetic neutral atoms) ribbon, ..."[47]

"The IBEX results are truly remarkable! What we are seeing in these maps does not match with any of the previous theoretical models of this region. It will be exciting for scientists to review these (ENA) maps and revise the way we understand our heliosphere and how it interacts with the galaxy."[48] In October 2010, significant changes were detected in the ribbon after 6 months, based on the second set of IBEX observations.[49] IBEX data did not support the existence of a bow shock,[6] but there might be a 'bow wave' according to one study.[7]

24.5.2 Locally

Of particular interest is the Earth's interaction with the heliosphere, but its extent and interaction with other bodies in the solar system has also been studied. Some examples of missions that have or continue to collect data related to the helopsphere include *(see also List of heliophysics missions)*:

- Solar Anomalous and Magnetospheric Particle Explorer]

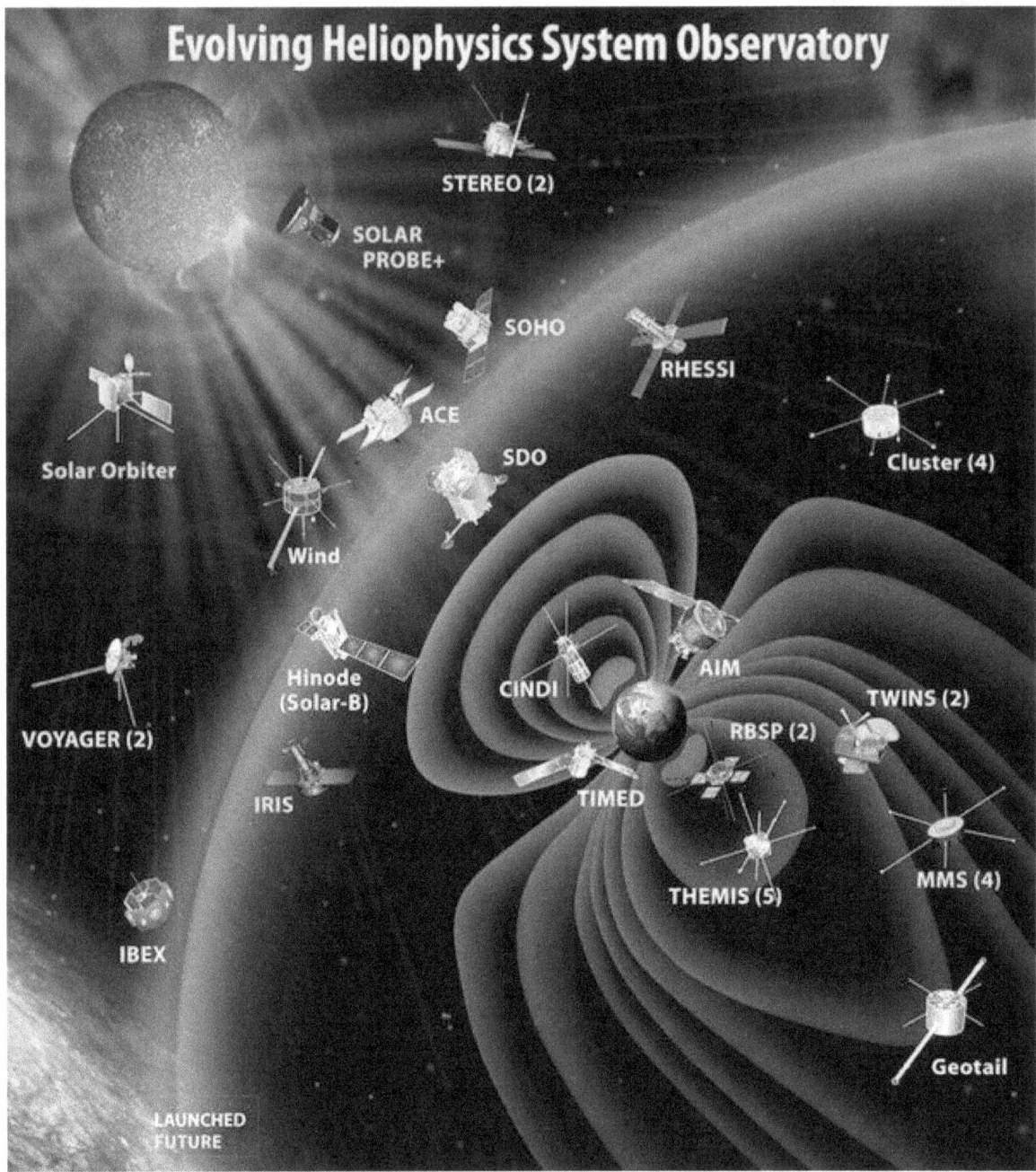

Overview of Heliophysics spacecraft circa 2011

- Solar and Heliosphere Observatory

- Solar Dynamics Observatory

- STEREO

- Ulysses (spacecraft)

During a total eclipse the high-temperature corona can be more readily observed from Earth solar observatories. During the Apollo program the Solar wind was measured on the Moon via the Solar Wind Composition Experiment. Some examples of Earth surface based Solar observatories include the McMath–Pierce solar telescope or the newer GREGOR Solar Telescope, and the refurbished Big Bear Solar Observatory.

24.5.3 Timeline

- As of March 2005, it was reported that measurements by the Solar Wind Anisotropies (SWAN) instrument on board the Solar and Heliospheric Observatory (SOHO) have shown that the heliosphere, the solar wind-filled volume which prevents the Solar System from becoming embedded in the local (ambient) interstellar medium, is not axisymmetrical, but is distorted, very likely under the effect of the local galactic magnetic field.[50]

- As of 2008, there is a previously unpredicted narrow ribbon of ENAs.[13]

- As of October 2009, the heliosphere may be bubble, not comet shaped.[12]

- As of October 2010, significant changes were detected in the ribbon after 6 months, based on the second set of IBEX observations.[49]

- As of June 2011, the heliosheath area is thought to be filled with magnetic bubbles (each about 1 AU wide), creating a "foamy zone".[16] The theory helps explain in situ heliosphere measurements by the two Voyager probes.

- As of May 2012, IBEX data implies there is probably not a bow "shock".[6]

- As of June 2012 at 119 AU Voyager 1 detected an increase in cosmic rays.[10]

- Between late August and early September 2012, Voyager 1 witnessed a sharp drop in protons from the sun, from 25 particles per sec in late August, to about 2 particles per second by early October.[51] It was later determined that it entered interstellar space on August 25, 2012.[3]

24.6 Gallery

These depictions include features that may be not reflect most recent models.[6][12][13][16]

- Heliosphere: classical model

- Heliosphere interacting with the interstellar wind

- Moving heliosphere, showing a heliosheath filled with magnetic bubble "foam" (red)[1]

- Play media

   Fragment of the film *Sentinels of the Heliosphere*, tracking some researching satellites deployed to analyse the Sun

1. ^ Cite error: The named reference surprise was invoked but never defined (see the help page).

24.7 See also

- Coronal mass ejection

- Fermi glow

- Heliospheric current sheet

- Interplanetary medium

- Interstellar medium

- Interstellar Boundary Explorer (IBEX)

- Pioneer 10 & Pioneer 11

- Solar flare

- Solar wind

- Tom Krimigis

- Ulysses probe

- Voyager Program

24.8   References

[1] Voyager 2 Proves Solar System Is Squashed NASA.gov #2007-12-10

[2] J. Matson - Voyager 1's Whereabouts: No News, but Plenty of Noise (2013) - Scientific American

[3] NASA Spacecraft Embarks on Historic Journey Into Interstellar Space (Sept. 2013)

[4] The Solar Wind

[5] NASA Voyager 1 Encounters New Region in Deep Space

[6] "New Interstellar Boundary Explorer data show heliosphere's long-theorized bow shock does not exist", *Phys.org*, May 10, 2012, retrieved 2012-02-11

[7] G. P. Zank, et al. - HELIOSPHERIC STRUCTURE: THE BOW WAVE AND THE HYDROGEN WALL (2013)

[8] NASA's Voyager Hits New Region at Solar System Edge 12.05.11

[9] NASA 2011

[10] NASA = Data From NASA's Voyager 1 Point to Interstellar Future 06.14.12

[11] NASA Spacecraft Embarks on Historic Journey Into Interstellar Space

[12] Johns Hopkins University (October 18, 2009). "New View Of The Heliosphere: Cassini Helps Redraw Shape Of Solar System". *ScienceDaily*. Retrieved October 22, 2009.

[13] "First IBEX Maps Reveal Fascinating Interactions Occurring At The Edge Of The Solar System".

[14] NASA's IBEX Provides First View Of the Solar System's Tail'

[15] Mursula, K.; Hiltula, T., (2003). "Bashful ballerina: Southward shifted heliospheric current sheet". *Geophysical Research Letters* **30** (22): 2135. Bibcode:2003GeoRL..30vSSC2M. doi:10.1029/2003GL018201.

[16] NASA - A Big Surprise from the Edge of the Solar System (06.09.11)

[17] Nemiroff, R.; Bonnell, J. (June 24, 2002). "The Sun's Heliosphere & Heliopause". Astronomy Picture of the Day. Retrieved 2007-05-25.

[18] "MIT instrument finds surprises at solar system's edge". Massachusetts Institute of Technology. 2007-12-10. Retrieved 2010-08-20.

[19] Steigerwald, Bill (May 24, 2005). "Voyager Enters Solar System's Final Frontier". American Astronomical Society. Retrieved 2007-05-25.

[20] "Voyager 2 Proves Solar System Is Squashed". Jet Propulsion Laboratory. December 10, 2007. Retrieved 2007-05-25.

[21] Donald A. Gurnett (1 June 2005). "Voyager Termination Shock". Department of Physics and Astronomy (University of Iowa). Retrieved 2008-02-06.

[22] Celeste Biever (25 May 2005). "Voyager 1 reaches the edge of the solar system". NewScientist. Retrieved 2008-02-06.

[23] David Shiga (10 December 2007). "Voyager 2 probe reaches solar system boundary". NewScientist. Retrieved 2008-02-06.

[24] Than, Ker (May 24, 2006). "Voyager II detects solar system's edge". CNN. Retrieved 2007-05-25.

[25] NASA - Transitional Regions at the Heliosphere's Outer Limits

[26] Brandt, Pontus (February 27 – March 2, 2007). "Imaging of the Heliospheric Boundary" (PDF). *NASA Advisory Council Workshop on Science Associated with the Lunar Exploration Architecture: White Papers*. Tempe, Arizona: Lunar and Planetary Institute. Retrieved 2007-05-25.

[27] Amos, Jonathan (December 14, 2010). "Voyager near Solar Systems edge". *BBC News*. Retrieved 2010-12-10.

[28] "NASA's Voyager 1 Spacecraft Nearing Edge of the Solar System". *Space.Com web site*. 2010-12-13. Retrieved 2010-12-15. External link in |work= (help)

[29] Brumfiel, G. (2011-06-15). "Voyager at the edge: spacecraft finds unexpected calm at the boundary of Sun's bubble". *Nature News web site*. doi:10.1038/news.2011.370. Retrieved 2011-06-19. External link in |work= (help)

[30] Krimigis, S. M.; Roelof, E. C.; Decker, R. B.; Hill, M. E. (2011-06-16). "Zero outward flow velocity for plasma in a heliosheath transition layer". *Nature* **474** (7351): 359–361. Bibcode:2011Natur.474..359K. doi:10.1038/nature10115. PMID 21677754. Retrieved 2011-06-20.

[31] Cook, J.-R. (2011-06-09). "NASA Probes Suggest Magnetic Bubbles Reside At Solar System Edge". NASA/JPL. Retrieved 2011-06-10.

[32] Rayl, A. j. s. (2011-06-12). "Voyager Discovers Possible Sea of Huge, Turbulent, Magnetic Bubbles at Solar System Edge". *The Planetary Society web site*. The Planetary Society. Retrieved 2011-06-13. External link in |work= (help)

[33] Cowen, R. (2013). "Voyager 1 has reached interstellar space". *Nature*. doi:10.1038/nature.2013.13735.

[34] Vergano, Dan. "Voyager 1 Leaves Solar System, NASA Confirms". *http://news.nationalgeographic.com/*. National Geographic. Retrieved 9 February 2015.

[35] The Unexpected Structure of the Heliotail, Astrobiology Magazine, July 12, 2013

[36] Cole, Steve, *NASA Satellite Provides First View of the Solar System's Tail,* NASA News Release 12-211, July 10, 2013

[37] Voyager Glossary

[38] NASA Spacecraft Embarks on Historic Journey Into Interstellar Space - Sept 12, 2013

[39] How Do We Know When Voyager Reaches Interstellar Space?

[40] Eleven Spacecraft Show Interstellar Wind Changed Direction Over 40 Years - Sept 5, 2013

[41] Wood, B. E.; Alexander, W. R.; Linsky, J. L. (July 13, 2006). "The Properties of the Local Interstellar Medium and the Interaction of the Stellar Winds of \epsilon Indi and \lambda Andromedae with the Interstellar Environment". American Astronomical Society. Retrieved 2007-05-25.

[42] Palmer, Jason (October 15, 2009). "BBC News article". Retrieved May 4, 2010.

[43] No Shocks for This Bow: IBEX Says We're Wrong

[44] "Pioneer H, Jupiter Swingby Out-of-the-Ecliptic Mission Study" (PDF). 20 August 1971. Retrieved 2 May 2012.

[45] NASA. "Revised no tail model of the heliosphere".

[46] Oct.15/09 IBEX team announcement at http://ibex.swri.edu/

[47] Kerr, Richard A. (2009). "Tying Up the Solar System With a Ribbon of Charged Particles". *Science* **326** (5951): 350–351. doi:10.1126/science.326_350a. PMID 19833930.

[48] Dave McComas, IBEX Principal Investigator at http://ibex.swri.edu/

[49] **The Ever-Changing Edge of the Solar System** (Oct/02/2010) - Astrobiology Magazine

[50] Lallement, R.; Quémerais, E.; Bertaux, J. L.; Ferron, S.; Koutroumpa, D.; Pellinen, R. (2005). "Deflection of the Interstellar Neutral Hydrogen Flow Across the Heliospheric Interface". *Science* **307** (5714): 1447–1449. Bibcode:2005Sci...307.1447L. doi:10.1126/science.1107953. PMID 15746421. Retrieved 2007-05-25.

[51] NBCnews.com. "Voyager spacecraft to leave solar system". Retrieved 2012-10-11.

24.8.1 Notes

- "Heliopause Seems to Be 23 Billion Kilometres". Universe Today. December 9, 2003. Retrieved 2007-08-08.

- "Space probes reveal Solar System's bullet shape". COSMOS magazine. May 11, 2007. Retrieved 2007-05-12.

24.9 External links

- Moving into Interstellar Space (Artist Concept)

- "Cassini Data Helps Redraw Shape of Our Solar System" 2010

- Publications in Refereed Journals

- Voyager Interstellar Mission Objectives

- The Heliosphere (Cosmicopia)

- NASA GALEX (Galaxy evolution Explorer) homepage at Caltech

- The Solar and Heliospheric Research Group at the University of Michigan

- Ribbon at Edge of Our Solar System: Will the Sun Enter a Million-Degree Cloud of Interstellar Gas this century ?

- A Big Surprise from the Edge of the Solar System (NASA 06.09.11)

- Schwadron, N. A.; et al. (6 September 2011). "Does the Space Environment Affect the Ecosphere?" (PDF). *Eos* (American Geophysical Union) **92** (36): 297–301. Bibcode:2011EOSTr..92..297S. doi:10.1029/2011eo360001.

24.9.1 Old theories

Warning, content may be outdated

- The Heliosphere, MIT Space Plasma Group

- UI's Don Gurnett Says Voyager 1 Is Approaching Edge Of Solar System December 8, 2003 Univ. of Iowa Press release

- NASA's Interstellar Probe (2000)

- CNN: NASA: Voyager I enters solar system's final frontier – May 25, 2005

- *New Scientist*: Voyager 1 reaches the edge of the solar system – May 25, 2005

- Surprises from the Edge of the Solar System – Voyager 1 Newest Findings as of September 2006

- The heliospheric hydrogen wall and astrospheres

- Heliosphere, has a diagram.

- Heliosphere Astronomy Cast episode #65, includes full transcript.

Chapter 25

Oort cloud

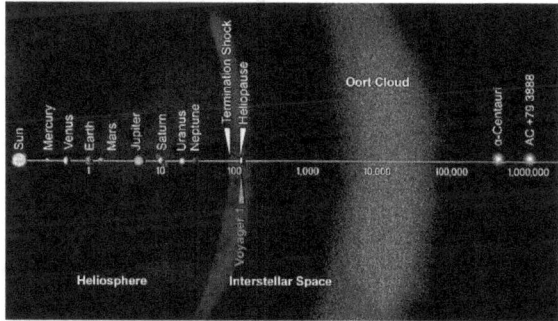

This graphic shows the distance from the Oort cloud to the rest of the Solar System and two of the nearest stars measured in astronomical units. The scale is logarithmic, with each specified distance ten times further out than the previous one.

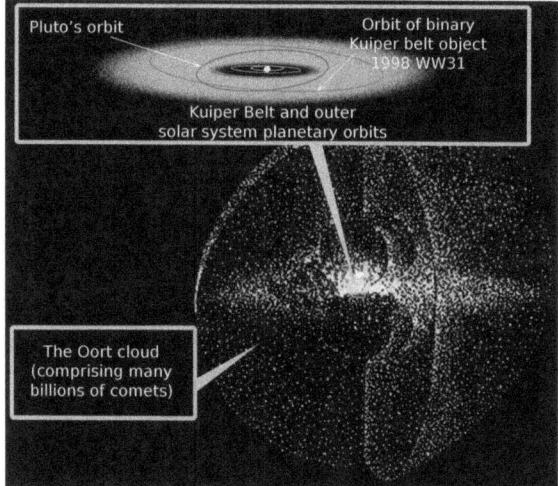

An artist's rendering of the Oort cloud and the Kuiper belt (inset). Sizes of individual objects have been exaggerated for visibility.

The **Oort cloud** (/'ɔrt/ or /'ʊərt/[1]) or **Öpik–Oort cloud**,[2] named after Dutch astronomer Jan Oort and Estonian astronomer Ernst Öpik, is a theoretical spherical cloud of predominantly icy planetesimals believed to surround the Sun at a distance of up to around 100,000 AU (2 ly).[3] This places it at almost half of the distance to Proxima Centauri, the nearest star to the Sun, and in interstellar space.[4] The Kuiper belt and the scattered disc, the other two reservoirs of trans-Neptunian objects, are less than one thousandth as far from the Sun as the Oort cloud. The outer limit of the Oort cloud defines the cosmographical boundary of the Solar System and the region of the Sun's gravitational dominance.[5]

The Oort cloud is thought to comprise two regions: a spherical outer Oort cloud and a disc-shaped inner Oort cloud, or **Hills cloud**. Objects in the Oort cloud are largely composed of ices, such as water, ammonia, and methane.

Astronomers conjecture that the matter composing the Oort cloud formed closer to the Sun and was scattered far into space by the gravitational effects of the giant planets early in the Solar System's evolution.[3] Although no confirmed direct observations of the Oort cloud have been made, it may be the source of all long-period and Halley-type comets entering the inner Solar System, and many of the centaurs and Jupiter-family comets as well.[6] The outer Oort cloud is only loosely bound to the Solar System, and thus is easily affected by the gravitational pull both of passing stars and of the Milky Way itself. These forces occasionally dislodge comets from their orbits within the cloud and send them towards the inner Solar System.[3] Based on their orbits, most of the short-period comets may come from the scattered disc, but some may still have originated from the Oort cloud.[3][6]

25.1   Hypothesis

In 1932, the Estonian astronomer Ernst Öpik postulated that long-period comets originated in an orbiting cloud at the outermost edge of the Solar System.[7] The idea was independently revived by Oort as a means to resolve a paradox.[8] Over the course of the Solar System's existence the orbits of comets are unstable and eventually dynamics dictate that a comet must either collide with the Sun or a planet or else be ejected from the Solar System by planetary perturbations. Moreover, their volatile composition means that as they repeatedly approach the Sun, radiation gradually boils the volatiles off until the comet splits or develops an insulating crust that prevents further outgassing. Thus, Oort reasoned, a comet could not have formed while in its current orbit and must have been held in an outer reservoir for almost all of its existence.[8][9][10]

There are two main classes of comet, short-period comets (also called ecliptic comets) and long-period comets (also called nearly isotropic comets). Ecliptic comets have relatively small orbits, below 10 AU, and follow the ecliptic plane, the same plane in which the planets lie. All long-period comets have very large orbits, on the order of thousands of AU, and appear from every direction in the sky.[10] Oort noted that there was a peak in numbers of long-period comets with aphelia (their farthest distance from the Sun) of roughly 20,000 AU, which suggested a reservoir at that distance with a spherical, isotropic distribution.[10] Those relatively rare comets with orbits of about 10,000 AU have probably gone through one or more orbits through the Solar System and have had their orbits drawn inward by the gravity of the planets.[10]

25.2   Structure and composition

The Oort cloud is thought to occupy a vast space from somewhere between 2,000 and 5,000 AU (0.03 and 0.08 ly)[10] to as far as 50,000 AU (0.79 ly)[3] from the Sun. Some estimates place the outer edge at between 100,000 and 200,000 AU (1.58 and 3.16 ly).[10] The region can be subdivided into a spherical outer Oort cloud of 20,000–50,000 AU (0.32–0.79 ly), and a doughnut-shaped inner Oort cloud of 2,000–20,000 AU (0.03–0.32 ly). The outer cloud is only weakly bound to the Sun and supplies the long-period (and possibly Halley-type) comets to inside the orbit of Neptune.[3] The inner Oort cloud is also known as the Hills cloud, named after Jack G. Hills, who proposed its existence in 1981.[11] Models predict that the inner cloud should have tens or hundreds of times as many cometary nuclei as the outer halo;[11][12][13] it is seen as a possible source of new comets to resupply the tenuous outer cloud as the latter's numbers are gradually depleted. The Hills cloud explains the continued existence of the Oort cloud after billions of years.[14]

The outer Oort cloud may have trillions of objects larger than 1 km (0.62 mi),[3] and billions with absolute magnitudes[15] brighter than 11 (corresponding to approximately 20-kilometre (12 mi) diameter), with neighboring objects tens of millions of kilometres apart.[6][16] Its total mass is not known, but, assuming that Halley's Comet is a suitable prototype for comets within the outer Oort cloud, roughly the combined mass is 3×10^{25} kilograms (6.6×10^{25} pounds), or five times that of Earth.[3][17] Earlier it was thought to be more massive (up to 380 Earth masses),[18] but improved knowledge of the size distribution of long-period comets led to lower estimates. The mass of the inner Oort cloud has not been characterized.

If analyses of comets are representative of the whole, the vast majority of Oort-cloud objects consist of ices such as water, methane, ethane, carbon monoxide and hydrogen cyanide.[19] However, the discovery of the object 1996 PW, an object whose appearance was consistent with a D-type asteroid[20][21] in an orbit typical of a long-period comet, prompted

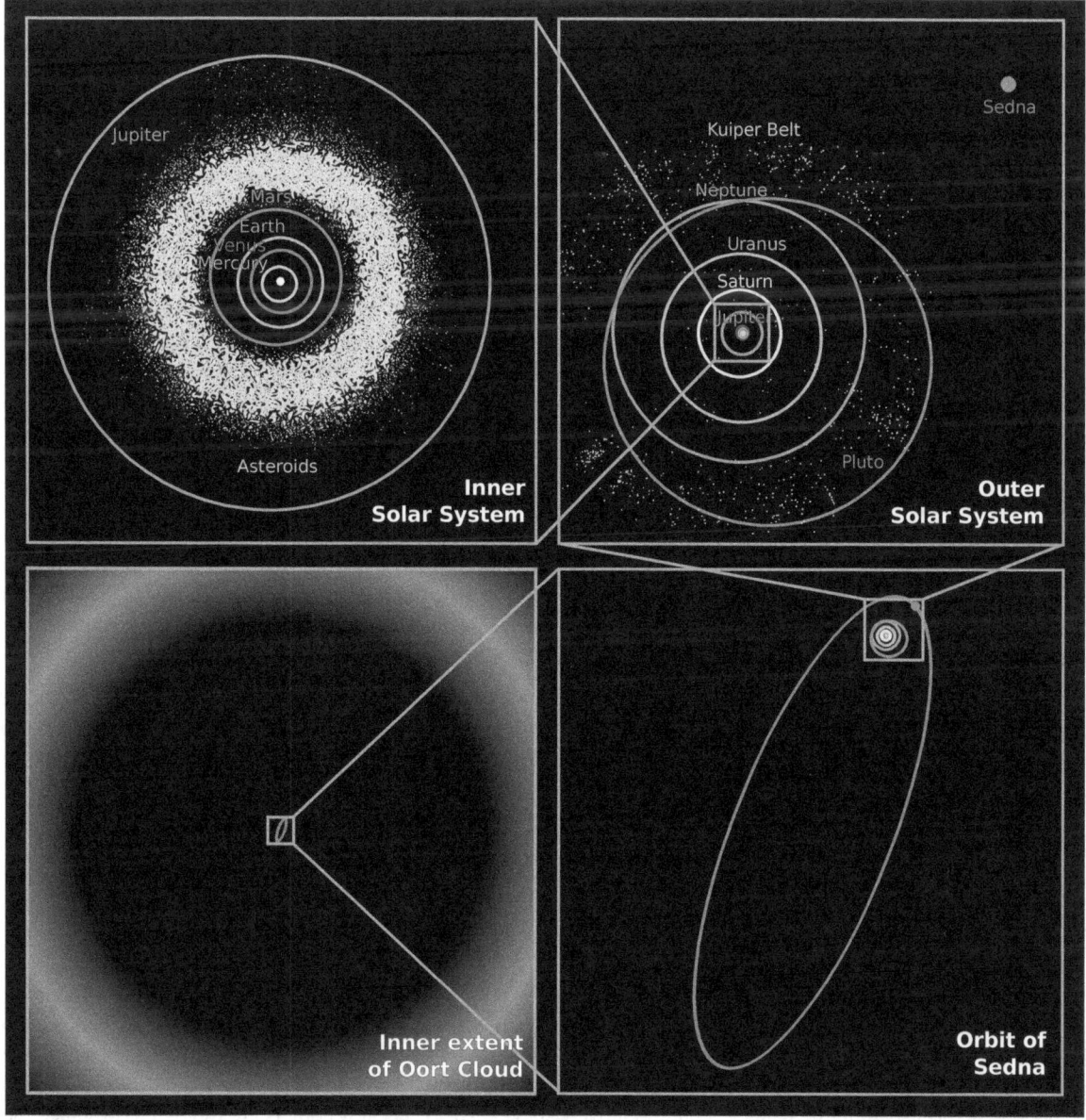

The presumed distance of the Oort cloud compared to the rest of the Solar System

theoretical research that suggests that the Oort cloud population consists of roughly one to two percent asteroids.[22] Analysis of the carbon and nitrogen isotope ratios in both the long-period and Jupiter-family comets shows little difference between the two, despite their presumably vastly separate regions of origin. This suggests that both originated from the original protosolar cloud,[23] a conclusion also supported by studies of granular size in Oort-cloud comets[24] and by the recent impact study of Jupiter-family comet Tempel 1.[25]

25.3   Origin

The Oort cloud is thought to be a remnant of the original protoplanetary disc that formed around the Sun approximately 4.6 billion years ago.[3] The most widely accepted hypothesis is that the Oort cloud's objects initially coalesced much closer to the Sun as part of the same process that formed the planets and minor planets, but that gravitational interaction with young gas giants such as Jupiter ejected the objects into extremely long elliptic or parabolic orbits.[3][26] Recent research has been cited by NASA hypothesizing that a large number of Oort cloud objects are the product of an exchange of materials between the Sun and its sibling stars as they formed and drifted apart, and it is suggested that many—possibly the majority—of Oort cloud objects were not formed in close proximity to the Sun.[27] Simulations of the evolution of the Oort cloud from the beginnings of the Solar System to the present suggest that the cloud's mass peaked around 800 million years after formation, as the pace of accretion and collision slowed and depletion began to overtake supply.[3]

Models by Julio Ángel Fernández suggest that the scattered disc, which is the main source for periodic comets in the Solar System, might also be the primary source for Oort cloud objects. According to the models, about half of the objects scattered travel outward towards the Oort cloud, whereas a quarter are shifted inward to Jupiter's orbit, and a quarter are ejected on hyperbolic orbits. The scattered disc might still be supplying the Oort cloud with material.[28] A third of the scattered disc's population is likely to end up in the Oort cloud after 2.5 billion years.[29]

Computer models suggest that collisions of cometary debris during the formation period play a far greater role than was previously thought. According to these models, the number of collisions early in the Solar System's history was so great that most comets were destroyed before they reached the Oort cloud. Therefore, the current cumulative mass of the Oort cloud is far less than was once suspected.[30] The estimated mass of the cloud is only a small part of the 50–100 Earth masses of ejected material.[3]

Gravitational interaction with nearby stars and galactic tides modified cometary orbits to make them more circular. This explains the nearly spherical shape of the outer Oort cloud.[3] On the other hand, the Hills cloud, which is bound more strongly to the Sun, has not acquired a spherical shape. Recent studies have shown that the formation of the Oort cloud is broadly compatible with the hypothesis that the Solar System formed as part of an embedded cluster of 200–400 stars. These early stars likely played a role in the cloud's formation, since the number of close stellar passages within the cluster was much higher than today, leading to far more frequent perturbations.[31]

In June 2010 Harold F. Levison and others suggested on the basis of enhanced computer simulations that the Sun "captured comets from other stars while it was in its birth cluster". Their results imply that "a substantial fraction of the Oort cloud comets, perhaps exceeding 90%, are from the protoplanetary discs of other stars".[32][33]

25.4   Comets

Comets are thought to have two separate points of origin in the Solar System. Short-period comets (those with orbits of up to 200 years) are generally accepted to have emerged from either the Kuiper belt or the scattered disc, which are two linked flat discs of icy debris beyond Neptune's orbit at 30 AU and jointly extending out beyond 100 AU from the Sun. Long-period comets, such as comet Hale–Bopp, whose orbits last for thousands of years, are thought to originate in the Oort cloud. The orbits within the Kuiper belt are relatively stable, and so very few comets are thought to originate there. The scattered disc, however, is dynamically active, and is far more likely to be the place of origin for comets.[10] Comets pass from the scattered disc into the realm of the outer planets, becoming what are known as centaurs.[34] These centaurs are then sent farther inward to become the short-period comets.[35]

There are two main varieties of short-period comet: Jupiter-family comets (those with semi-major axes of less than 5 AU) and Halley-family comets. Halley-family comets, named for their prototype, Halley's Comet, are unusual in that although they are short-period comets, it is hypothesized that their ultimate origin lies in the Oort cloud, not in the scattered disc. Based on their orbits, it is suggested they were long-period comets that were captured by the gravity of the giant planets and sent into the inner Solar System.[9] This process may have also created the present orbits of a significant fraction of the Jupiter-family comets, although the majority of such comets are thought to have originated in the scattered disc.[6]

Oort noted that the number of returning comets was far less than his model predicted, and this issue, known as "cometary fading", has yet to be resolved. No known dynamical process can explain this undercount of observed comets. Hypotheses for this discrepancy include the destruction of comets due to tidal stresses, impact or heating; the loss of all volatiles, rendering some comets invisible, or the formation of a non-volatile crust on the surface.[36] Dynamical studies of Oort cloud comets have shown that their occurrence in the outer-planet region is several times higher than in the inner-planet region. This discrepancy may be due to the gravitational attraction of Jupiter, which acts as a kind of barrier, trapping incoming comets and causing them to collide with it, just as it did with Comet Shoemaker–Levy 9 in 1994.[37]

25.5 Tidal effects

Main article: Galactic tide

Most of the comets seen close to the Sun seem to have reached their current positions through gravitational perturbation of the Oort cloud by the tidal force exerted by the Milky Way. Just as the Moon's tidal force deforms Earth's oceans, causing the tides to rise and fall, the galactic tide also distorts the orbits of bodies in the outer Solar System. In the charted regions of the Solar System, these effects are negligible compared to the gravity of the Sun, but in the outer reaches of the system, the Sun's gravity is weaker and the gradient of the Milky Way's gravitational field has substantial effects. Galactic tidal forces stretch the cloud along an axis directed toward the galactic centre and compress it along the other two axes; these small perturbations can shift orbits in the Oort cloud to bring objects close to the Sun.[38] The point at which the Sun's gravity concedes its influence to the galactic tide is called the tidal truncation radius. It lies at a radius of 100,000 to 200,000 AU, and marks the outer boundary of the Oort cloud.[10]

Some scholars theorise that the galactic tide may have contributed to the formation of the Oort cloud by increasing the perihelia (smallest distances to the Sun) of planetesimals with large aphelia (largest distances to the Sun).[39] The effects of the galactic tide are quite complex, and depend heavily on the behaviour of individual objects within a planetary system. Cumulatively, however, the effect can be quite significant: up to 90% of all comets originating from the Oort cloud may be the result of the galactic tide.[40] Statistical models of the observed orbits of long-period comets argue that the galactic tide is the principal means by which their orbits are perturbed toward the inner Solar System.[41]

25.6 Stellar perturbations and stellar companion hypotheses

Besides the galactic tide, the main trigger for sending comets into the inner Solar System is thought to be interaction between the Sun's Oort cloud and the gravitational fields of nearby stars[3] or giant molecular clouds.[37] The orbit of the Sun through the plane of the Milky Way sometimes brings it in relatively close proximity to other stellar systems. For example, 70 thousand years ago, Scholz's star passed through the outer Oort cloud (although its low mass and high relative velocity limited its effect).[42] During the next 10 million years the known star with the greatest possibility of perturbing the Oort cloud is Gliese 710.[43] This process also scatters Oort cloud objects out of the ecliptic plane, potentially also explaining its spherical distribution.[43][44]

In 1984, Physicist Richard A. Muller postulated that the Sun has a heretofore undetected companion, either a brown dwarf or a red dwarf, in an elliptical orbit within the Oort cloud. This object, known as Nemesis, was hypothesized to pass through a portion of the Oort cloud approximately every 26 million years, bombarding the inner Solar System with comets. However, to date no evidence of Nemesis has been found, and many lines of evidence (such as crater counts), have thrown its existence into doubt.[45][46] Recent scientific analysis no longer supports the idea that extinctions on Earth happen at regular, repeating intervals.[47] Thus, the Nemesis hypothesis is no longer needed.[47]

A somewhat similar hypothesis was advanced by astronomer John J. Matese of the University of Louisiana at Lafayette in 2002. He contends that more comets are arriving in the inner Solar System from a particular region of the Oort cloud than can be explained by the galactic tide or stellar perturbations alone, and that the most likely cause is a Jupiter-mass object in a distant orbit.[48] This hypothetical gas giant was nicknamed Tyche. The WISE mission, an all-sky survey using parallax measurements in order to clarify local star distances, was capable of proving or disproving the Tyche hypothesis.[47] In 2014, NASA announced that the WISE survey had ruled out any object as they had defined it.[49]

25.7 Modified Newtonian dynamics within the Oort cloud

Modified Newtonian dynamics (MOND)[50][51] suggests that at their distances from the Sun, the objects comprising the Oort cloud should experience accelerations of the order of 10^{-10} m/s^2, and thus should be within the realms at which deviations from Newtonian predictions come into effect. According to this hypothesis, which was proposed to account for the discrepancies in the galaxy rotation curve, which are more commonly attributed to dark matter, acceleration ceases to be linearly proportional to force at very low accelerations.[50] If correct, this would have significant implications regarding the formation and structure of the Oort cloud. However, the majority of cosmologists do not consider MOND a valid hypothesis because it is unable to explain the movement of galactic clusters or account accurately for the cosmic microwave background.[52]

25.8 Future exploration

Space probes have yet to reach the area of the Oort cloud. *Voyager 1*, the fastest[53] and farthest[54][55] of the interplanetary space probes currently exiting the Solar System, will reach the Oort cloud in about 300 years[4][56] and would take about 30,000 years to pass through it.[57][58] However, around 2025, *Voyager 1*'s radioisotope thermoelectric generators will no longer supply enough power to operate any of its scientific instruments, preventing any meaningful exploration by *Voyager 1*. The other four probes currently escaping the Solar System either are already or are predicted to be non-functional when they reach the Oort cloud, however it may be possible find an object Oort from the cloud that has been knocked into the inner solar system.

One proposal for exploration is to use a craft powered by a solar sail that would take around 30 years to reach its destination.[59] In the 1980s there was a concept for probe to reach 1000 AU in 50 years called TAU, among its missions would be to look for the Oort cloud.[60]

In the 2014 AO for the Discovery program, a observatory to detect the objects in the Oort cloud (and Kupier Belt) called the "The Whipple Mission" was proposed.[61] It would monitor distant stars with photometer, looking for transits of out to 10 thousand AU.[62] The observatory was proposed for halo orbit around L2 with a suggest 5-year mission.[63] It has been suggest that Kepler observatory may also be able to detect objects in the Oort cloud.[64]

25.9 See also

- Heliosphere
- Interstellar comet
- Kuiper belt
- List of possible dwarf planets
- List of trans-Neptunian objects
- Scattered disc
- Tyche (hypothetical planet)

25.10 References

[1] "Oort". *Oxford English Dictionary* (3rd ed.). Oxford University Press. September 2005. (Subscription or UK public library membership required.)

[2] Whipple, F. L.; Turner, G.; McDonnell, J. A. M.; Wallis, M. K. (1987-09-30). "A Review of Cometary Sciences". *Philosophical Transactions of the Royal Society A* (Royal Society Publishing) **323** (1572): 339–347 [341]. Bibcode:1987RSPTA.323..339W. doi:10.1098/rsta.1987.0090.

[3] Alessandro Morbidelli (2006). "Origin and dynamical evolution of comets and their reservoirs of water ammonia and methane". arXiv:astro-ph/0512256 [astro-ph].

[4] "Catalog Page for PIA17046". *Photo Journal*. NASA. Retrieved April 27, 2014.

[5] "Kuiper Belt & Oort Cloud". *NASA Solar System Exploration web site*. NASA. Retrieved 2011-08-08. External link in |work= (help)

[6] V. V. Emelyanenko; D. J. Asher; M. E. Bailey (2007). "The fundamental role of the Oort Cloud in determining the flux of comets through the planetary system". *Monthly Notices of the Royal Astronomical Society* **381** (2): 779–789. Bibcode:2007MNRAS.381..779E. doi:10.1111/j.1365-2966.2007.12269.x.

[7] Ernst Julius Öpik (1932). "Note on Stellar Perturbations of Nearby Parabolic Orbits". *Proceedings of the American Academy of Arts and Sciences* **67** (6): 169–182. doi:10.2307/20022899. JSTOR 20022899.

[8] Jan Oort (1950). "The structure of the cloud of comets surrounding the Solar System and a hypothesis concerning its origin". *Bulletin of the Astronomical Institutes of the Netherlands* **11**: 91–110. Bibcode:1950BAN....11...91O.

[9] David C. Jewitt (2001). "From Kuiper Belt to Cometary Nucleus: The Missing Ultrared Matter". *Astronomical Journal* **123** (2): 1039–1049. Bibcode:2002AJ....123.1039J. doi:10.1086/338692.

[10] Harold F. Levison; Luke Donnes (2007). "Comet Populations and Cometary Dynamics". In Lucy Ann Adams McFadden; Lucy-Ann Adams; Paul Robert Weissman; Torrence V. Johnson. *Encyclopedia of the Solar System* (2nd ed.). Amsterdam; Boston: Academic Press. pp. 575–588. ISBN 0-12-088589-1.

[11] Jack G. Hills (1981). "Comet showers and the steady-state infall of comets from the Oort Cloud". *Astronomical Journal* **86**: 1730–1740. Bibcode:1981AJ.....86.1730H. doi:10.1086/113058.

[12] Harold F. Levison; Luke Dones; Martin J. Duncan (2001). "The Origin of Halley-Type Comets: Probing the Inner Oort Cloud". *Astronomical Journal* **121** (4): 2253–2267. Bibcode:2001AJ....121.2253L. doi:10.1086/319943.

[13] Thomas M. Donahue, ed. (1991). *Planetary Sciences: American and Soviet Research, Proceedings from the U.S.–U.S.S.R. Workshop on Planetary Sciences.* Kathleen Kearney Trivers, and David M. Abramson. National Academy Press. p. 251. ISBN 0-309-04333-6. Retrieved 2008-03-18.

[14] Julio A. Fernéndez (1997). "The Formation of the Oort Cloud and the Primitive Galactic Environment" (PDF). *Icarus* **219**: 106–119. Bibcode:1997Icar..129..106F. doi:10.1006/icar.1997.5754. Retrieved 2008-03-18.

[15] Absolute magnitude is a measure of how bright an object would be if it were 1 AU from the Sun and Earth; as opposed to apparent magnitude, which measures how bright an object appears from Earth. Because all measurements of absolute magnitude assume the same distance, absolute magnitude is in effect a measurement of an object's brightness. The lower an object's absolute magnitude, the brighter it is.

[16] Paul R. Weissman (1998). "The Oort Cloud". *Scientific American*. Retrieved 2007-05-26.

[17] Paul R. Weissman (1983). "The mass of the Oort Cloud". *Astronomy and Astrophysics* **118** (1): 90–94. Bibcode:1983A&A...118...90W.

[18] Sebastian Buhai. "On the Origin of the Long Period Comets: Competing theories" (PDF). Utrecht University College. Archived from the original (PDF) on 2006-09-30. Retrieved 2008-03-29.

[19] E. L. Gibb; M. J. Mumma; N. Dello Russo; M. A. DiSanti & K. Magee-Sauer (2003). "Methane in Oort Cloud comets". *Icarus* **165** (2): 391–406. Bibcode:2003Icar..165..391G. doi:10.1016/S0019-1035(03)00201-X.

[20] Rabinowitz, D. L. (August 1996). "1996 PW". *IAU circular* (International Astronomical Union) **6466**. Bibcode:1996IAUC.6466....2R.

[21] Davies, John K.; McBride, Neil; Green, Simon F.; Mottola, Stefano; et al. (April 1998). "The Lightcurve and Colors of Unusual Minor Planet 1996 PW". *Icarus* (Elsevier) **132** (2): 418–430. Bibcode:1998Icar..132..418D. doi:10.1006/icar.1998.5888. (subscription required (help)).

[22] Paul R. Weissman; Harold F. Levison (1997). "Origin and Evolution of the Unusual Object 1996 PW: Asteroids from the Oort Cloud?". *Astrophysical Journal* **488** (2): L133–L136. Bibcode:1997ApJ...488L.133W. doi:10.1086/310940.

[23] D. Hutsemekers; J. Manfroid; E. Jehin; C. Arpigny; A. Cochran; R. Schulz; J.A. Stüwe & J M. Zucconi (2005). "Isotopic abundances of carbon and nitrogen in Jupiter-family and Oort Cloud comets". *Astronomy and Astrophysics* **440** (2): L21–L24. arXiv:astro-ph/0508033. Bibcode:2005A&A...440L..21H. doi:10.1051/0004-6361:200500160.

[24] Takafumi Ootsubo; Jun-ichi Watanabe; Hideyo Kawakita; Mitsuhiko Honda & Reiko Furusho (2007). "Grain properties of Oort Cloud comets: Modeling the mineralogical composition of cometary dust from mid-infrared emission features". *Highlights in Planetary Science, 2nd General Assembly of Asia Oceania Geophysical Society* **55** (9): 1044–1049. Bibcode:2007P&SS...55.1044O. doi:10.1016/j.pss.2006.11.012.

[25] Michael J. Mumma; Michael A. DiSanti; Karen Magee-Sauer; et al. (2005). "Parent Volatiles in Comet 9P/Tempel 1: Before and After Impact". *Science Express* **310** (5746): 270–274. Bibcode:2005Sci...310..270M. doi:10.1126/science.1119337. PMID 16166477.

[26] "Oort Cloud & Sol b?". SolStation. Retrieved 2007-05-26.

[27] "The Sun Steals Comets from Other Stars". NASA. 2010.

[28] Julio A. Fernández; Tabaré Gallardo & Adrián Brunini (2004). "The scattered disc population as a source of Oort Cloud comets: evaluation of its current and past role in populating the Oort Cloud". *Icarus* **172** (2): 372–381. Bibcode:2004Icar..172..372F. doi:10.1016/j.icarus.2004.07.023.

[29] Davies, J. K.; Barrera, L. H. (2004). *The First Decadal Review of the Edgeworth-Kuiper Belt.* Kluwer Academic Publishers. ISBN 978-1-4020-1781-0.

[30] S. Alan Stern; Paul R. Weissman (2001). "Rapid collisional evolution of comets during the formation of the Oort Cloud". *Nature* **409** (6820): 589–591. Bibcode:2001Natur.409..589S. doi:10.1038/35054508. PMID 11214311.

[31] R. Brasser; M. J. Duncan; H.F. Levison (2006). "Embedded star clusters and the formation of the Oort Cloud". *Icarus* **184** (1): 59–82. Bibcode:2006Icar..184...59B. doi:10.1016/j.icarus.2006.04.010.

[32] Levison, Harold; et al. (10 June 2010). "Capture of the Sun's Oort Cloud from Stars in Its Birth Cluster". *Science* **329** (5988): 187–190. Bibcode:2010Sci...329..187L. doi:10.1126/science.1187535.

[33] "Many famous comets originally formed in other solar systems". *Southwest Research Institute® (SwRI®) News.* 10 June 2010.

[34] Harold E. Levison & Luke Dones (2007). "Comet Populations and Cometary dynamics". *Encyclopedia of the Solar System*: 575–588. doi:10.1016/B978-012088589-3/50035-9. ISBN 978-0-12-088589-3.

[35] J Horner; NW Evans; ME Bailey; DJ Asher (2003). "The Populations of Comet-like Bodies in the Solar System" (PDF). Retrieved 2007-06-29.

[36] Luke Dones; Paul R Weissman; Harold F Levison; Martin J Duncan (2004). "Oort Cloud Formation and Dynamics" (PDF). In Michel C. Festou; H. Uwe Keller; Harold A. Weaver. *Comets II*. University of Arizona Press. pp. 153–173. Retrieved 2008-03-22.

[37] Julio A. Fernández (2000). "Long-Period Comets and the Oort Cloud". *Earth, Moon, and Planets* **89** (1–4): 325–343. Bibcode:2002EM&P...89..325F. doi:10.1023/A:1021571108658.

[38] Marc Fouchard; Christiane Froeschlé; Giovanni Valsecchi; Hans Rickman (2006). "Long-term effects of the galactic tide on cometary dynamics". *Celestial Mechanics and Dynamical Astronomy* **95** (1–4): 299–326. Bibcode:2006CeMDA..95..299F. doi:10.1007/s10569-006-9027-8.

[39] Higuchi A.; Kokubo E. & Mukai, T. (2005). "Orbital Evolution of Planetesimals by the Galactic Tide". *Bulletin of the American Astronomical Society* **37**: 521. Bibcode:2005DDA....36.0205H.

[40] Nurmi P.; Valtonen M.J.; Zheng J.Q. (2001). "Periodic variation of Oort Cloud flux and cometary impacts on the Earth and Jupiter". *Monthly Notices of the Royal Astronomical Society* **327** (4): 1367–1376. Bibcode:2001MNRAS.327.1367N. doi:10.1046/j.1365-8711.2001.04854.x.

[41] John J. Matese & Jack J. Lissauer (2004). "Perihelion evolution of observed new comets implies the dominance of the galactic tide in making Oort Cloud comets discernible". *Icarus* **170** (2): 508–513. Bibcode:2004Icar..170..508M. doi:10.1016/j.icarus.2004.03.019.

[42] Mamajek, Eric E.; Barenfeld, Scott A.; Ivanov, Valentin D. (2015). "The Closest Known Flyby of a Star to the Solar System". *The Astrophysical Journal* **800** (1). arXiv:1502.04655. Bibcode:2015ApJ...800L..17M. doi:10.1088/2041-8205/800/1/L17.

[43] L. A. Molnar; R. L. Mutel (1997). *Close Approaches of Stars to the Oort Cloud: Algol and Gliese 710.* American Astronomical Society 191st meeting. American Astronomical Society. Bibcode:1997AAS...191.6906M.

[44] A. Higuchi; E. Kokubo & T. Mukai (2006). "Scattering of Planetesimals by a Planet: Formation of Comet Cloud Candidates". *Astronomical Journal* **131** (2): 1119–1129. Bibcode:2006AJ....131.1119H. doi:10.1086/498892.

[45] J. G. Hills (1984). "Dynamical constraints on the mass and perihelion distance of Nemesis and the stability of its orbit". *Nature* **311** (5987): 636–638. Bibcode:1984Natur.311..636H. doi:10.1038/311636a0.

[46] "Nemesis is a myth". Max Planck Institute. 2011. Retrieved 2011-08-11.

[47] "Can WISE Find the Hypothetical 'Tyche'?". NASA/JPL. February 18, 2011. Retrieved 2011-06-15.

[48] John J. Matese & Jack J. Lissauer (2002-05-06). "Continuing Evidence of an Impulsive Component of Oort Cloud Cometary Flux" (PDF). University of Louisiana at Lafayette, and NASA Ames Research Center. Retrieved 2008-03-21.

[49] K. L., Luhman (7 March 2014). "A Search For A Distant Companion To The Sun With The Wide-field Infrared Survey Explorer". *The Astrophysical Journal* **781** (1). Bibcode:2014ApJ...781....4L. doi:10.1088/0004-637X/781/1/4. Retrieved 20 March 2014.

[50] Milgrom, M. (1983). "A modification of the newtonian dynamics as a possible alternative to the hidden mass hypothesis". *Astrophysical Journal* **270**: 365–370. Bibcode:1983ApJ...270..365M. doi:10.1086/161130.

[51] Milgrom, M. (1986). "Solutions for the modified Newtonian dynamics field equation". *Astrophysical Journal* **302**: 617–625. Bibcode:1986ApJ...302..617M. doi:10.1086/164021.

[52] Sean Carroll. "Dark Matter: Just Fine, Thanks". *Discover*. Retrieved 2011-03-04.

[53] "New Horizons Salutes Voyager". New Horizons. August 17, 2006. Retrieved November 3, 2009.

[54] Clark, Stuart (September 13, 2013). "Voyager 1 leaving solar system matches feats of great human explorers". *The Guardian*.

[55] "Voyagers are leaving the Solar System". *Space Today*. 2011. Retrieved May 29, 2014.

[56] "It's Official: Voyager 1 Is Now In Interstellar Space". *UniverseToday*. Retrieved April 27, 2014.

[57] Ghose, Tia (September 13, 2013). "Voyager 1 Really Is In Interstellar Space: How NASA Knows". *Space.com web site*. TechMedia Network. Retrieved September 14, 2013. External link in |work= (help)

[58] Cook, J.-R (September 12, 2013). "How Do We Know When Voyager Reaches Interstellar Space?". NASA / Jet Propulsion Lab. Retrieved September 15, 2013.

[59] Paul Gilster (2008-11-12). "An Inflatable Sail to the Oort Cloud". Centauri-dreams.org. Retrieved 2013-07-23.

[60]

[61]

[62]

[63]

[64] Scientific American - Kepler Spacecraft May Be Able to Spot Elusive Oort Cloud Objects - 2010

25.11 External links

- Oort Cloud Profile by NASA's Solar System Exploration

- The Kuiper Belt and The Oort Cloud

- The effect of perturbations by the Alpha Cen A/B system on the Oort Cloud

- Reassessing the formation of the Inner Oort cloud in an embedded star cluster II: Probing the inner edge (Brasser; Schwamb : 7 Nov 2014 : arXiv:1411.1844)

Comet Hale–Bopp, an archetypical Oort-cloud comet

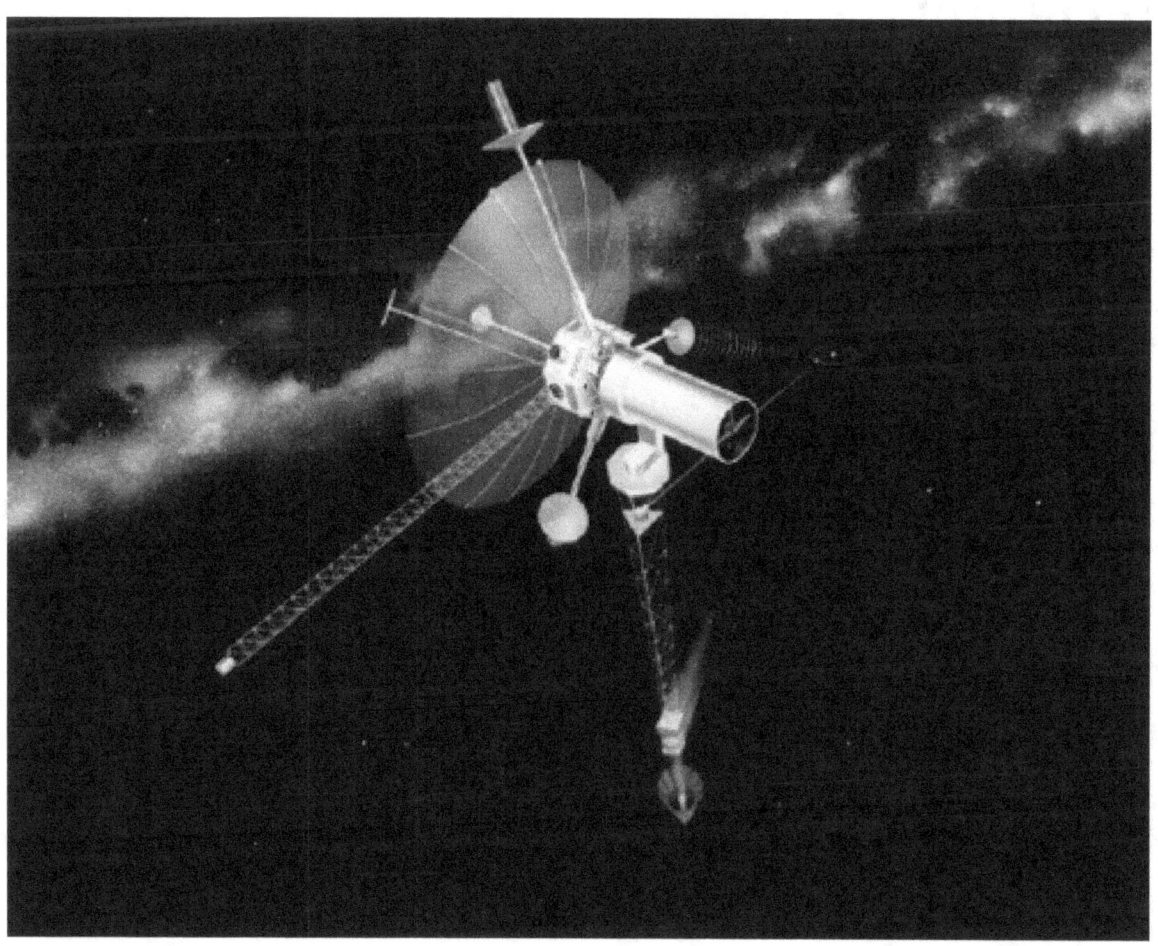

TAU concept art

Chapter 26

Oort limit

The Oort limit is a theoretical location at the outer limits of the Oort cloud, where the amount of comets and minor planets orbiting the Sun drops drastically, or drops entirely. The exact location of such a limit, if there is such one, is uncertain. About 100 comets, of 3500 known comets, come more than 5000 AU from the Sun, and a very few come as far as 20,000 AU from the Sun. So far, it appears that rather than a sudden drop in the amount of comets orbiting the sun at about 50,000 AU, the Oort cloud rather uniformly decreases in size, the further away from the Sun it goes. As current observations indicate, the Oort limit is somewhere around 50,000 AU (0.8 LY) from the Sun.

26.1   See also

^ Of these known comets, the majority (>2000) were discovered using the SOHO telescope, and are mostly sungrazing comets from the Kreutz Sungrazers. Of the other comets, about half of them are long-period comets, orbiting several hundred Astronomical Units out or further. Considering this, Oort Cloud comets are fairly common.

26.2   See also

- Kuiper Cliff

26.3   References

- The Encyclopedia of Astrobiology, Astronomy, and Spaceflight

Chapter 27

Scattered disc

The **scattered disc** (or **scattered disk**) is a distant region of the Solar System that is sparsely populated by icy minor planets, a subset of the broader family of trans-Neptunian objects. The scattered-disc objects (SDOs) have orbital eccentricities ranging as high as 0.8, inclinations as high as 40°, and perihelia greater than 30 astronomical units (4.5×10^9 km; 2.8×10^9 mi). These extreme orbits are thought to be the result of gravitational "scattering" by the gas giants, and the objects continue to be subject to perturbation by the planet Neptune.

Although the closest scattered-disc objects approach the Sun at about 30–35 AU, their orbits can extend well beyond 100 AU. This makes scattered objects among the most distant and coldest objects in the Solar System.[1] The innermost portion of the scattered disc overlaps with a torus-shaped region of orbiting objects traditionally called the Kuiper belt,[2] but its outer limits reach much farther away from the Sun and farther above and below the ecliptic than the Kuiper belt proper.[lower-alpha 1]

Because of its unstable nature, astronomers now consider the scattered disc to be the place of origin for most periodic comets in the Solar System, with the centaurs, a population of icy bodies between Jupiter and Neptune, being the intermediate stage in an object's migration from the disc to the inner Solar System.[4] Eventually, perturbations from the giant planets send such objects towards the Sun, transforming them into periodic comets. Many Oort cloud objects are also thought to have originated in the scattered disc. Detached objects are not sharply distinct from scattered disc objects, and some such as Sedna have sometimes been considered to be included in this group.

27.1 Discovery

See also: History of the Kuiper belt

Traditionally, devices like a blink comparator were used in astronomy to detect objects in the Solar System, because these objects would move between two exposures—this involved time-consuming steps like exposing and developing photographic plates or films, and people then using a blink comparator to manually detect prospective objects. During the 1980s, the use of CCD-based cameras in telescopes made it possible to directly produce electronic images that could then be readily digitized and transferred to digital images. Because the CCD captured more light than film (about 90% versus 10% of incoming light) and the blinking could now be done at an adjustable computer screen, the surveys allowed for higher throughput. A flood of new discoveries was the result: over a thousand trans-Neptunian objects were detected between 1992 and 2006.[5]

The first scattered-disc object (SDO) to be recognised as such was 1996 TL_{66},[6][7] originally identified in 1996 by astronomers based at Mauna Kea in Hawaii. Three more were identified by the same survey in 1999: 1999 CV_{118}, 1999 CY_{118}, and 1999 CF_{119}.[8] The first object presently classified as an SDO to be discovered was 1995 TL_8, found in 1995 by Spacewatch.[9]

As of 2011, over 200 SDOs have been identified,[10] including 2007 UK_{126} (discovered by Schwamb, Brown, and

Eris, the largest known scattered-disc object (center), and its moon Dysnomia (left of object)

Rabinowitz),[11] 2002 TC$_{302}$ (NEAT), Eris (Brown, Trujillo, and Rabinowitz),[12] Sedna (Brown, Trujillo, and Rabinowitz)[13] and 2004 VN112 (Deep Ecliptic Survey).[14] Although the numbers of objects in the Kuiper belt and the scattered disc are hypothesized to be roughly equal, observational bias due to their greater distance means that far fewer SDOs have been observed to date.[15]

27.2   Subdivisions of trans-Neptunian space

Main article: Trans-Neptunian object

 Known trans-Neptunian objects are often divided into two subpopulations: the Kuiper belt and the scattered disc.[16] A third reservoir of trans-Neptunian objects, the Oort cloud, has been hypothesized, although no confirmed direct observations of the Oort cloud have been made.[2] Some researchers further suggest a transitional space between the scattered disc and the inner Oort cloud, populated with "detached objects".[17]

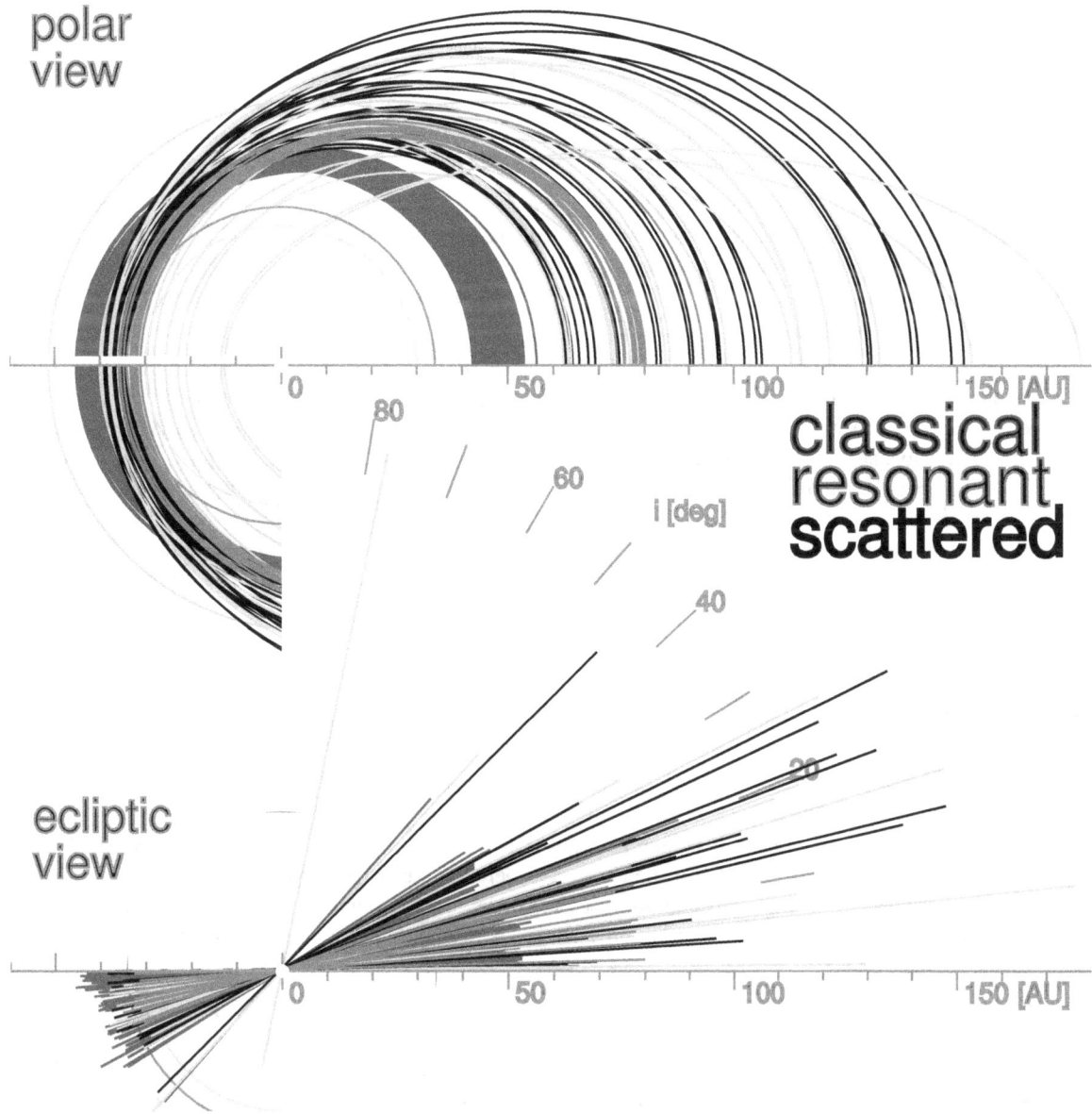

The eccentricity and inclination of the scattered-disc population compared to the classical and 5:2 resonant Kuiper-belt objects

27.2.1 Scattered disc versus Kuiper belt

See also: Kuiper belt

The Kuiper belt is a relatively thick torus (or "doughnut") of space, extending from about 30 to 50 AU[18] comprising two main populations of Kuiper belt objects (KBOs): the classical Kuiper-belt objects (or "cubewanos"), which lie in orbits untouched by Neptune, and the resonant Kuiper-belt objects; those which Neptune has locked into a precise orbital ratio such as 3:2 (the object goes around twice for every three Neptune orbits) and 2:1 (the object goes around once for every two Neptune orbits). These ratios, called orbital resonances, allow KBOs to persist in regions which Neptune's gravitational influence would otherwise have cleared out over the age of the Solar System, since the objects are never close enough to Neptune to be scattered by its gravity. Those in 3:2 resonances are known as "plutinos", because Pluto is the largest member of their group, whereas those in 2:1 resonances are known as "twotinos".

In contrast to the Kuiper belt, the scattered-disc population can be disturbed by Neptune.[19] Scattered-disc objects come within gravitational range of Neptune at their closest approaches (~30 AU) but their farthest distances reach many times that.[17] Ongoing research[20] suggests that the centaurs, a class of icy planetoids that orbit between Jupiter and Neptune, may simply be SDOs thrown into the inner reaches of the Solar System by Neptune, making them "cis-Neptunian" rather than trans-Neptunian scattered objects.[21] Some objects, like (29981) 1999 TD_{10}, blur the distinction[22] and the Minor Planet Center (MPC), which officially catalogues all trans-Neptunian objects, now lists centaurs and SDOs together.[10]

The MPC also makes a clear distinction between the Kuiper belt and the scattered disc; separating those objects in stable orbits (the Kuiper belt) from those in scattered orbits (the scattered disc and the centaurs).[10] However, the difference between the Kuiper belt and the scattered disc is not clearcut, and many astronomers see the scattered disc not as a separate population but as an outward region of the Kuiper belt. Another term used is "scattered Kuiper-belt object" (or SKBO) for bodies of the scattered disc.[23]

Morbidelli and Brown propose that the difference between objects in the Kuiper-belt and scattered-disc objects is that the latter bodies "are transported in semi-major axis by close and distant encounters with Neptune",[16] but the former experienced no such close encounters. This delineation is inadequate (as they note) over the age of the Solar System, since bodies "trapped in resonances" could "pass from a scattering phase to a non-scattering phase (and vice versa) numerous times".[16] That is, trans-Neptunian objects could travel back and forth between the Kuiper belt and the scattered disc over time. Therefore they chose instead to define the regions, rather than the objects, defining the scattered disc as "the region of orbital space that can be visited by bodies that have encountered Neptune" within the radius of a Hill sphere, and the Kuiper belt as its "complement ... in the $a > 30$ AU region"; the region of the Solar System populated by objects with semi-major axes greater than 30 AU.[16]

27.2.2   Detached objects

Main article: Detached object

The Minor Planet Center classifies the trans-Neptunian object 90377 Sedna as a scattered-disc object. Its discoverer Michael E. Brown has suggested instead that it should be considered an inner Oort-cloud object rather than a member of the scattered disc, because, with a perihelion distance of 76 AU, it is too remote to be affected by the gravitational attraction of the outer planets. [24] Under this definition, an object with a perihelion greater than 40 AU could be classified as outside the scattered disc. [25]

Sedna is not the only such object: 2000 CR105 (discovered before Sedna) and 2004 VN112 have a perihelion too far away from Neptune to be influenced by it. This led to a discussion among astronomers about a new minor planet set, called the *extended scattered disc* (**E-SDO**). [26] 2000 CR105 may also be an inner Oort-cloud object or (more likely) a transitional object between the scattered disc and the inner Oort cloud. More recently, these objects have been referred to as *"detached"*,[27] or *distant detached objects* (**DDO**).[28]

There are no clear boundaries between the scattered and detached regions.[25] Gomes et al. define SDOs as having "highly eccentric orbits, perihelia beyond Neptune, and semi-major axes beyond the 1:2 resonance." By this definition, all distant detached objects are SDOs.[17] Since detached objects' orbits cannot be produced by Neptune scattering, alternative scattering mechanisms have been put forward, including a passing star[29] or a distant, planet-sized object.[28]

A scheme introduced by a 2005 report from the Deep Ecliptic Survey by J. L. Elliott et al. distinguishes between two categories: *scattered-near* (i.e. typical SDOs) and *scattered-extended* (i.e. detached objects).[30] Scattered-near objects are those whose orbits are non-resonant, non-planetary-orbit-crossing and have a Tisserand parameter (relative to Neptune) less than 3.[30] Scattered-extended objects have a Tisserand parameter (relative to Neptune) greater than 3 and have a time-averaged eccentricity greater than 0.2.[30]

An alternative classification, introduced by B. Gladman, B. Marsden and C. VanLaerhoven in 2007, uses 10-million-year orbit integration instead of the Tisserand parameter.[31] An object qualifies as an SDO if its orbit is not resonant, has a semi-major axis no greater than 2000 AU, and, during the integration, its semi-major axis shows an excursion of 1.5 AU or more.[31] Gladman et al. suggest the term *scattering disk object* to emphasize this present mobility.[31] If the object is not an SDO as per the above definition, but the eccentricity of its orbit is greater than 0.240, it is classified as a *detached TNO*.[31] (Objects with smaller eccentricity are considered classical.) In this scheme, the disc extends from the orbit of

Neptune to 2000 AU, the region referred to as the inner Oort cloud.

27.3   Orbits

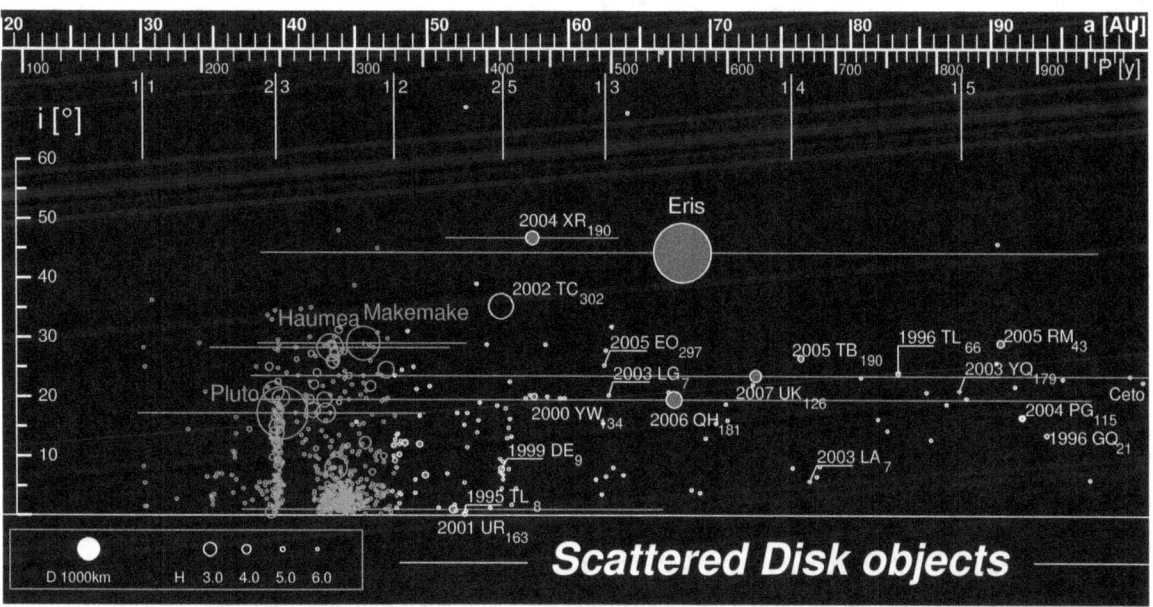

The semi-major axes and inclinations of all known scattered-disc objects (in blue) up to 100 AU together with Kuiper-belt objects (in grey) and resonant objects (in green). The eccentricity of the orbits is represented by segments (extending from the perihelion to the aphelion) with the inclination represented on Y axis.

The scattered disc is a very dynamic environment.[15] Because they are still capable of being perturbed by Neptune, SDOs' orbits are always in danger of disruption; either of being sent outward to the Oort cloud or inward into the centaur population and ultimately the Jupiter family of comets.[15] For this reason Gladman et al. prefer to refer to the region as the scattering disc, rather than scattered.[31] Unlike Kuiper-belt objects (KBOs), the orbits of scattered-disc objects can be inclined as much as 40° from the ecliptic.[32]

SDOs are typically characterized by orbits with medium and high eccentricities with a semi-major axis greater than 50 AU, but their perihelia bring them within influence of Neptune.[33] Having a perihelion of roughly 30 AU is one of the defining characteristics of scattered objects, as it allows Neptune to exert its gravitational influence.[8]

The classical objects (cubewanos) are very different from the scattered objects: more than 30% of all cubewanos are on low-inclination, near-circular orbits whose eccentricities peak at 0.25.[34] Classical objects possess eccentricities ranging from 0.2 to 0.8. Though the inclinations of scattered objects are similar to the more extreme KBOs, very few scattered objects have orbits as close to the ecliptic as much of the KBO population.[15]

Although motions in the scattered disc are random, they do tend to follow similar directions, which means that SDOs can become trapped in temporary resonances with Neptune. Examples of resonant orbits within the scattered disc include 1:3, 2:7, 3:11, 5:22 and 4:79.[17]

27.4   Formation

See also: Formation and evolution of the Solar System

 The scattered disc is still poorly understood: no model of the formation of the Kuiper belt and the scattered disc has yet been proposed that explains all their observed properties.[16]

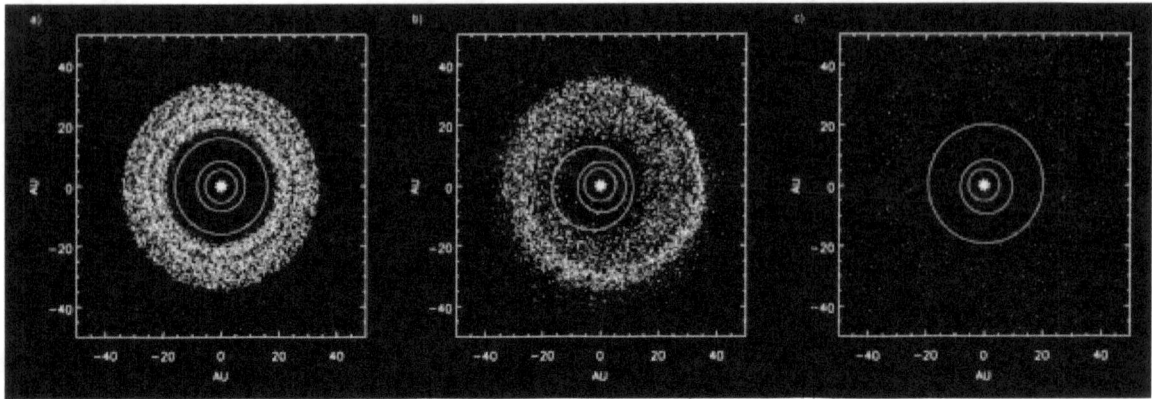

Simulation showing Outer Planets and Kuiper Belt: a) Before Jupiter/Saturn 2:1 resonance b) Scattering of Kuiper-belt objects into the Solar System after the orbital shift of Neptune c) After ejection of Kuiper-belt bodies by Jupiter

According to contemporary models, the scattered disc formed when Kuiper belt objects (KBOs) were "scattered" into eccentric and inclined orbits by gravitational interaction with Neptune and the other outer planets.[35] The amount of time for this process to occur remains uncertain. One hypothesis estimates a period equal to the entire age of the Solar System;[36] a second posits that the scattering took place relatively quickly, during Neptune's early migration epoch.[37]

Models for a continuous formation throughout the age of the Solar System illustrate that at weak resonances within the Kuiper belt (such as 5:7 or 8:1), or at the boundaries of stronger resonances, objects can develop weak orbital instabilities over millions of years. The 4:7 resonance in particular has large instability. KBOs can also be shifted into unstable orbits by close passage of massive objects, or through collisions. Over time, the scattered disc would gradually form from these isolated events.[17]

Computer simulations have also suggested a more rapid and earlier formation for the scattered disc. Modern theories indicate that neither Uranus nor Neptune could have formed *in situ* beyond Saturn, as too little primordial matter existed at that range to produce objects of such high mass. Instead, these planets, and Saturn, may have formed closer to Jupiter, but were flung outwards during the early evolution of the Solar System, perhaps through exchanges of angular momentum with scattered objects.[38] Once the orbits of Jupiter and Saturn shifted to a 2:1 resonance (two Jupiter orbits for each orbit of Saturn), their combined gravitational pull disrupted the orbits of Uranus and Neptune, sending Neptune into the temporary "chaos" of the proto-Kuiper belt.[37] As Neptune traveled outward, it scattered many trans-Neptunian objects into higher and more eccentric orbits.[35][39] This model states that 90% or more of the objects in the scattered disc may have been "promoted into these eccentric orbits by Neptune's resonances during the migration epoch...[therefore] the scattered disc might not be so scattered."[40]

27.5 Composition

Scattered objects, like other trans-Neptunian objects, have low densities and are composed largely of frozen volatiles such as water and methane.[41] Spectral analysis of selected Kuiper belt and scattered objects has revealed signatures of similar compounds. Both Pluto and Eris, for instance, show signatures for methane.[42]

Astronomers originally supposed that the entire trans-Neptunian population would show a similar red surface colour, as they were thought to have originated in the same region and subjected to the same physical processes.[41] Specifically, SDOs were expected to have large amounts of surface methane, chemically altered into complex organic molecules by energy from the Sun. This would absorb blue light, creating a reddish hue.[41] Most classical objects display this colour, but scattered objects do not; instead, they present a white or greyish appearance.[41]

One explanation is the exposure of whiter subsurface layers by impacts; another is that the scattered objects' greater distance from the Sun creates a composition gradient, analogous to the composition gradient of the terrestrial and gas giant planets.[41] Mike Brown, discoverer of the scattered object Eris, suggests that its paler colour could be because, at its current distance from the Sun, its atmosphere of methane is frozen over its entire surface, creating an inches-thick layer

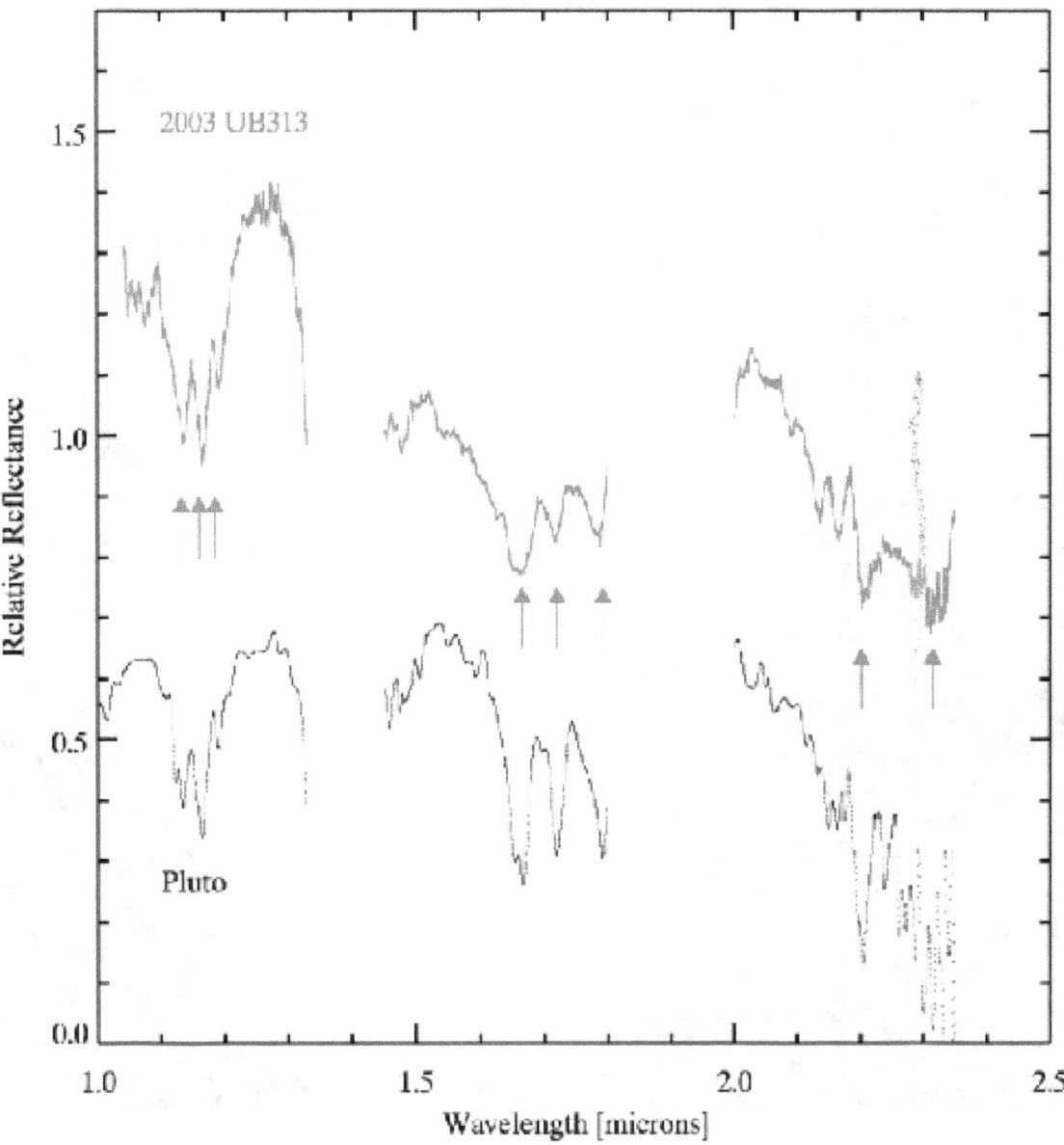

The infrared spectra of both Eris and Pluto, highlighting their common methane absorption lines

of bright white ice. Pluto, conversely, being closer to the Sun, would be warm enough that methane would freeze only onto cooler, high-albedo regions, leaving low-albedo tholin-covered regions bare of ice.[42]

27.6 Comets

Main article: Comet § Short period

The Kuiper belt was initially thought to be the source of the Solar System's ecliptic comets. However, studies of the region since 1992 have shown that the orbits within the Kuiper belt are relatively stable, and that these comets originate from the scattered disc, where orbits are generally less stable.[43]

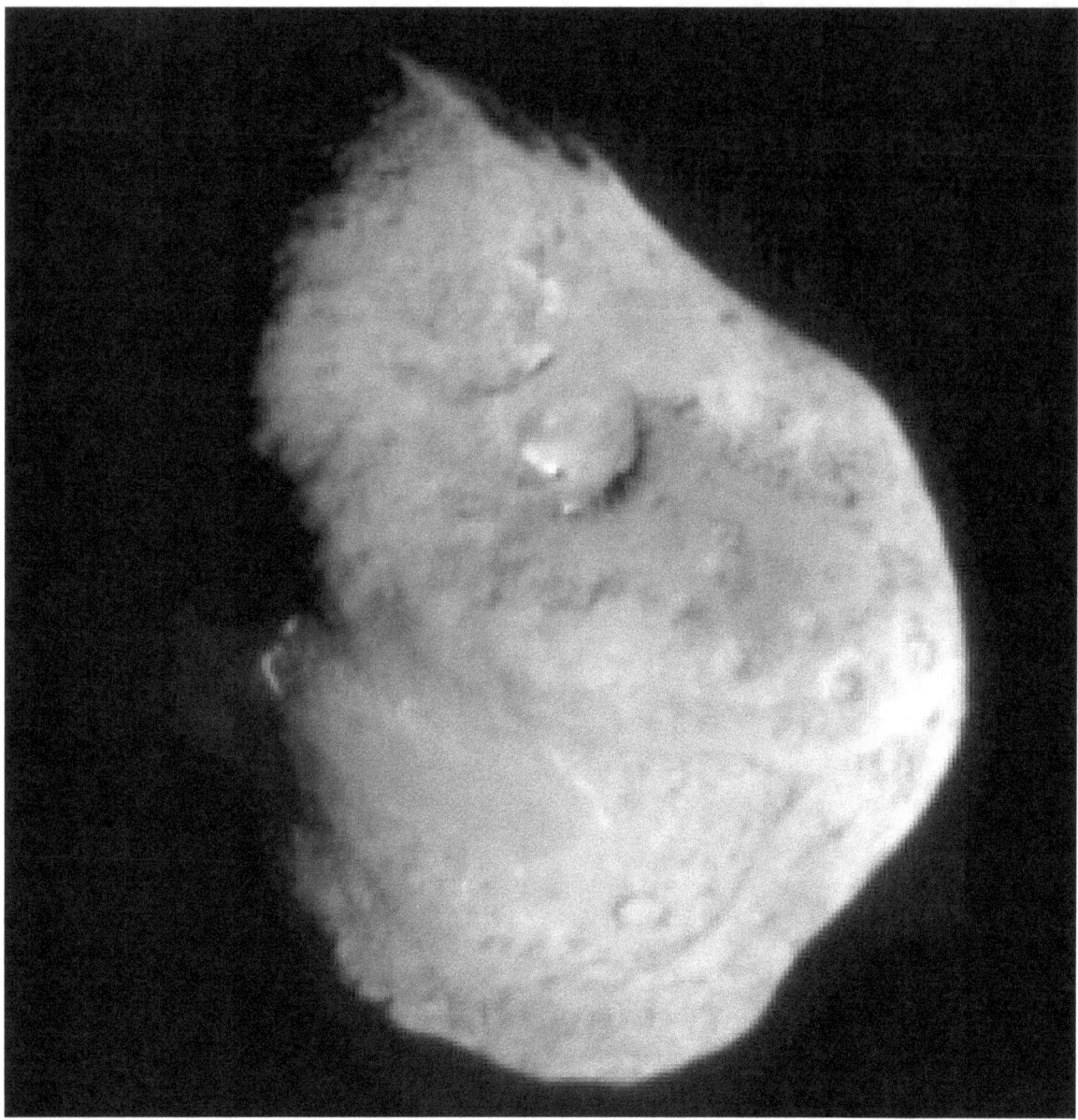

Tempel 1, a Jupiter-family comet

Comets can loosely be divided into two categories: short-period and long-period—the latter being thought to originate in the Oort cloud. The two major categories of short-period comets are Jupiter-family comets (JFCs) and Halley-type comets.[15] Halley-type comets, which are named after their prototype, Halley's Comet, are thought to have originated in the Oort cloud but to have been drawn into the inner Solar System by the gravity of the giant planets,[44] whereas the JFCs are thought to have originated in the scattered disc.[19] The centaurs are thought to be a dynamically intermediate stage between the scattered disc and the Jupiter family.[20]

There are many differences between SDOs and JFCs, even though many of the Jupiter-family comets may have originated in the scattered disc. Although the centaurs share a reddish or neutral coloration with many SDOs, their nuclei are bluer, indicating a fundamental chemical or physical difference.[44] One hypothesis is that comet nuclei are resurfaced as they approach the Sun by subsurface materials which subsequently bury the older material.[44]

27.7   See also

- List of possible dwarf planets

- List of trans-Neptunian objects

27.8   Notes

[1] The literature is inconsistent in the use of the phrases "scattered disc" and "Kuiper belt". For some, they are distinct populations; for others, the scattered disc is part of the Kuiper belt. Authors may even switch between these two uses in a single publication.[3] In this article, the scattered disc will be considered a separate population from the Kuiper belt.

27.9   References

[1] Maggie Masetti. (2007). *Cosmic Distance Scales – The Solar System*. Website of NASA's High Energy Astrophysics Science Archive Research Center. Retrieved 2008 07-12.

[2] Alessandro Morbidelli (2005). "Origin and dynamical evolution of comets and their reservoirs". arXiv:astro-ph/0512256 [astro-ph].

[3] McFadden, Weissman, & Johnson (2007). *Encyclopedia of the Solar System*, footnote p. 584

[4] Horner, J.; Evans, N.W.; Bailey, M. E. (2004). "Simulations of the Population of Centaurs I: The Bulk Statistics". *Monthly Notices of the Royal Astronomical Society* **354** (3): 798. arXiv:astro-ph/0407400. Bibcode:2004MNRAS.354..798H. doi:10.1111/j.1365-2966.2004.08240.x.

[5] Scott S Sheppard (October 16–18, 2005). "Small Bodies in the Outer Solar System" (PDF). *New Horizons in Astronomy: Frank N. Bash Symposium 2005*. Austin, Texas: Astronomical Society of the Pacific. pp. 3–14. ISBN 1-58381-220-2. Retrieved 2008-08-14.

[6] Jane Luu; Brian G. Marsden; David Jewitt; et al. (5 June 1997). "A new dynamical class of object in the outer Solar System" (PDF). *Nature* **387** (6633): 573–575. Bibcode:1997Natur.387..573L. doi:10.1038/42413. Archived from the original (PDF) on August 12, 2007. Retrieved 2008-08-02.

[7] John Keith Davies (2001). *Beyond Pluto: Exploring the Outer Limits of the Solar System* (PDF). Cambridge University Press. p. 111. ISBN 0-521-80019-6. Retrieved 2008-07-02.

[8] David Jewitt (August 2009). "Scattered Kuiper Belt Objects (SKBOs)". Institute for Astronomy. Retrieved 2010-01-23.

[9] Lutz D. Schmadel, (2003). *Dictionary of Minor Planet Names* (5th rev. and enlarged ed. edition). Berlin: Springer. Page 925 (Appendix 10). Also see McFadden, Lucy-Ann, Weissman, Paul & Johnson, Torrence (1999). *Encyclopedia of the Solar System*. San Diego: Academic Press. Page 218.

[10] IAU: Minor Planet Center (2011-01-03). "List Of Centaurs and Scattered-Disk Objects". Central Bureau for Astronomical Telegrams, Harvard-Smithsonian Center for Astrophysics. Retrieved 2011-01-03.

[11] Schwamb, M. E.; Brown, M. E.; Rabinowitz, D.; Marsden, B. G. (2008). "2007 UK126". *Minor Planet Electronic Circ.*: 38. Bibcode:2008MPEC....D...38S.

[12] Staff (2007-05-01). "Discovery Circumstances: Numbered Minor Planets". Minor Planet Center. Retrieved 2010-10-25.

[13] "Discovery Circumstances: Numbered Minor Planets (90001)-(95000)". Minor Planet Center. Retrieved 2010-10-25.

[14] Marc W. Buie (2007-11-08). "Orbit Fit and Astrometric record for 04VN112". SwRI (Space Science Department). Retrieved 2008-07-17.

[15] Harold F. Levison; Luke Donnes (2007). "Comet Populations and Cometary Dynamics". In Lucy Ann Adams McFadden; Lucy-Ann Adams; Paul Robert Weissman; Torrence V. Johnson. *Encyclopedia of the Solar System* (2nd ed.). Amsterdam; Boston: Academic Press. pp. 575–588. ISBN 0-12-088589-1.

[16] Alessandro Morbidelli; ME Brown (2004-11-01). "The Kuiper Belt and the Primordial Evolution of the Solar System". In MC Festou; HU Keller; HA Weaver. *Comets II* (PDF). Tucson: University of Arizona Press. pp. 175–91. ISBN 0-8165-2450-5. OCLC 56755773. Retrieved 2008-07-27.

[17] Rodney S. Gomes; Julio A. Fernandez; Tabare Gallardo; Adrian Brunini (2008). "The Scattered Disk: Origins, Dynamics and End States" (PDF). *Universidad de la Republica, Uruguay.* Retrieved 2008-08-10.

[18] M. C. De Sanctis; M. T. Capria & A. Coradini (2001). "Thermal Evolution and Differentiation of Edgeworth-Kuiper Belt Objects". *The Astronomical Journal* **121** (5): 2792–2799. Bibcode:2001AJ....121.2792D. doi:10.1086/320385.

[19] Alessandro Morbidelli; Harold F. Levison (2007). "Kuiper Belt Dynamics". In Lucy-Ann Adams McFadden; Paul Robert Weissman; Torrence V. Johnson. *Encyclopedia of the Solar System* (2nd ed.). Amsterdam; Boston: Academic Press. pp. 589–604. ISBN 0-12-088589-1.

[20] J Horner; NW Evans; ME Bailey; DJ Asher (2003). "The Populations of Comet-like Bodies in the Solar System" (PDF). *Monthly Notices of the Royal Astronomical Society* **343** (4): 1057–1066. arXiv:astro-ph/0304319. Bibcode:2003MNRAS.343.1057H. doi:10.1046/j.1365-8711.2003.06714.x. Retrieved 2007-06-29.

[21] Remo notes that Cis-Neptunian bodies "include terrestrial and large gaseous planets, planetary moons, asteroids, and main-belt comets within Neptune's orbit."(Remo 2007)

[22] Kenneth Silber (1999). "New Object in Solar System Defies Categories". *space.com*. Retrieved 2008-08-12.

[23] David Jewitt (2008). "The 1000 km Scale KBOs". Retrieved 2010-01-23.

[24] Michael E. Brown. "Sedna (The coldest most distant place known in the solar system; possibly the first object in the long-hypothesized Oort cloud)". California Institute of Technology, Department of Geological Sciences. Retrieved 2008-07-02.

[25] Patryk Sofia Lykawka; Tadashi Mukai (2007). "Dynamical classification of trans-Neptunian objects: Probing their origin, evolution, and interrelation". *Icarus* (Kobe) **189** (1): 213–232. Bibcode:2007Icar..189..213L. doi:10.1016/j.icarus.2007.01.001. Retrieved 2008-07-24.

[26] Brett Gladman. "Evidence for an Extended Scattered Disk?". *Observatoire de la Cote d'Azur.* Retrieved 2008-08-02.

[27] David C. Jewitt; A. Delsanti (2006). "The Solar System Beyond The Planets". *Solar System Update : Topical and Timely Reviews in Solar System Sciences.* Springer-Praxis Ed. ISBN 3-540-26056-0. (Preprint version (pdf))

[28] Rodney S Gomes; John J. Matese; Jack J. Lissauer (October 2006). "A distant planetary-mass solar companion may have produced distant detached objects". *Icarus* **184** (2): 589–601. Bibcode:2006Icar..184..589G. doi:10.1016/j.icarus.2006.05.026.

[29] Alessandro Morbidelli; Harold F. Levison (November 2004). "Scenarios for the Origin of the Orbits of the Trans-Neptunian Objects 2000 CR105 and 2003 VB12". *The Astronomical Journal* **128** (5): 2564–2576. arXiv:astro-ph/0403358. Bibcode:2004AJ....128.2564M. doi:10.1086/424617. Retrieved 2008-07-02.

[30] J. L. Elliot; S. D. Kern; K. B. Clancy; et al. (2005). "The Deep Ecliptic Survey: A Search for Kuiper Belt Objects and Centaurs. II. Dynamical Classification, the Kuiper Belt Plane, and the Core Population" (PDF). *The Astronomical Journal* **129** (2): 1117–1162. Bibcode:2005AJ....129.1117E. doi:10.1086/427395. Archived (PDF) from the original on June 25, 2008.

[31] B. Gladman; B. Marsden; C. VanLaerhoven (2008). "Nomenclature in the Outer Solar System". *In The Solar System Beyond Neptune, ISBN 978-0-8165-2755-7*: 43. Bibcode:2008ssbn.book...43G.

[32] Bertoldi, F.; Altenhoff, W.; Weiss, A.; Menten, K. M.; Thum, C. (2 February 2006). "The trans-Neptunian object UB$_{313}$ is larger than Pluto". *Nature* **439** (7076): 563–564. Bibcode:2006Natur.439..563B. doi:10.1038/nature04494. PMID 16452973.

[33] Chadwick A. Trujillo; David C. Jewitt; Jane X. Luu (2000-02-01). "Population of the Scattered Kuiper Belt" (PDF). *The Astrophysical Journal* **529** (2): L103–L106. arXiv:astro-ph/9912428. Bibcode:2000ApJ...529L.103T. doi:10.1086/312467. PMID 10622765. Archived from the original (PDF) on August 12, 2007. Retrieved 2008-07-02.

[34] Levison, H. F.; Morbidelli, A. (2003-11-27). "The formation of the Kuiper belt by the outward transport of bodies during Neptune's migration". *Nature* **426** (6965): 419–421. Bibcode:2003Natur.426..419L. doi:10.1038/nature02120. PMID 14647375. Retrieved 2012-05-26.

[35] Duncan, Martin J.; Levison, Harold F. (1997). "A Disk of Scattered Icy Objects and the Origin of Jupiter-Family Comets". *Science* **276** (5319): 1670–1672. Bibcode:1997Sci...276.1670D. doi:10.1126/science.276.5319.1670. PMID 9180070.

[36] Harold F. Levison; Martin J Duncan (1997). "From the Kuiper Belt to Jupiter-Family Comets: The Spatial Distribution of Ecliptic Comets". *Icarus* **127** (1): 13–32. Bibcode:1997Icar..127...13L. doi:10.1006/icar.1996.5637. Retrieved 2008-07-18.

[37] Kathryn Hansen (2005-06-07). "Orbital shuffle for early solar system". *Geotimes*. Retrieved 2007-08-26.

[38] Joseph M. Hahn; Renu Malhotra (13 July 2005). "Neptune's Migration into a Stirred–Up Kuiper Belt: A Detailed Comparison of Simulations to Observations". *Astronomical Journal* **130** (5): 2392. arXiv:astro-ph/0507319. Bibcode:2005AJ....130.2392H. doi:10.1086/452638.

[39] E. W.Thommes; MJ Duncan; HF Levison (May 2002). "The Formation of Uranus and Neptune Among Jupiter and Saturn". *The Astronomical Journal* **123** (5): 2862–83. arXiv:astro-ph/0111290. Bibcode:2002AJ....123.2862T. doi:10.1086/339975.

[40] Joseph M Hahn; Renu Malhotra (November 2005). "Neptune's Migration into a Stirred-Up Kuiper Belt: A Detailed Comparison of Simulations to Observations". *The Astronomical Journal* **130** (5): 2392–414. arXiv:astro-ph/0507319. Bibcode:2005AJ....130.2392H. doi:10.1086/452638.

[41] Stephen C. Tegler (2007). "Kuiper Belt Objects: Physical Studies". In Lucy Ann Adams McFadden; Paul Robert Weissman; Torrence V. Johnson. *Encyclopedia of the Solar System* (2nd ed.). Amsterdam; Boston: Academic Press. pp. 605–620. ISBN 0-12-088589-1.

[42] Michael E. Brown; Chadwick A. Trujillo; David L. Rabinowitz (2005). "Discovery of a Planetary-sized Object in the Scattered Kuiper Belt". *The Astrophysical Journal* **635** (1): L97–L100. arXiv:astro-ph/0508633. Bibcode:2005ApJ...635L..97B. doi:10.1086/499336.

[43] Gladman, Brett (2005). "The Kuiper Belt and the Solar System's Comet Disk". *Science* **307** (5706): 71–75. Bibcode:2005Sci...307...71G. doi:10.1126/science.1100553. PMID 15637267.

[44] David C Jewitt (2001). "From Kuiper Belt Object to Cometary Nucleus: The Missing Ultrared Matter". *The Astronomical Journal* **123** (2): 1039–1049. Bibcode:2002AJ....123.1039J. doi:10.1086/338692.

Chapter 28

Taiwanese–American Occultation Survey

The **Taiwanese–American Occultation Survey** (**TAOS**) is a robotic survey of the Outer Solar System.[1][2] TAOS uses an array of four 50 cm aperture telescopes to monitor background stars awaiting the alignment of an Outer Solar System with a star target: an *occultation*. Small objects in the Outer Solar System that are too small to be observed by direct observations at this time can be probed with this technique. Occultation surveys take advantage of diffraction effects during the transit of the occulting object (the *occulter*) in front of a background star to constraint the size and distance of the occulter. TAOS is sensitive to occultations by Kuiper Belt Objects (**KBOs**) larger than about 500 m in diameter [3] and to Sedna-like objects.

The TAOS telescopes are located in Taiwan, at the Lulin Observatory in Yushan national park.

TAOS is a joint effort of the Academia Sinica Institute of Astronomy and Astrophysics, Harvard-Smithsonian Center for Astrophysics, Lawrence Livermore National Laboratory, The Institute of Geophysics and Planetary Physics, National Central University, Institute of Astronomy and Yonsei University, South Korea.

Currently, an expansion to the TAOS project is being planned called TAOS II, with a new meaning to the acronym, the Transneptunian Automated Occultation Survey. The data volume is expected to be over 300 terabytes per year.

28.1 References

[1] C. Alcock et al. (2004). "TAOS: The Taiwanese–American Occultation Survey". In John Keith Davies; Luis H. Barrera. *The First Decadal Review of the Edgeworth–Kuiper Belt*. Springer. pp. 459–464. ISBN 978-1-4020-1781-0.

[2] Lehner, M. J.; Wen, C.-Y.; Wang, J.-H.; Marshall, S. L.; Schwamb, M. E.; Zhang, Z.-W.; Bianco, F. B.; Giammarco, J.; Porrata, R.; Alcock, C.; Axelrod, T.; Byun, Y.-I.; Chen, W. P.; Cook, K. H.; Dave, R.; King, S.-K.; Lee, T.; Lin, H.-C.; Wang, S.-Y.; Rice, J. A.; de Pater, I.The Taiwanese-American Occultation Survey: The Multi-Telescope Robotic Observatory Publications of the Astronomical Society of the Pacific, Volume 121, issue 876, pp.138-152 (PASP Homepage) doi:10.1086/597516

[3] Z.-W. Zhang, F. B. Bianco, M. J. Lehner, N. K. Coehlo, J.-H. Wang, S. Mondal, C. Alcock, T. Axelrod, Y.-I. Byun, W.-P. Chen, K. H. Cook, R. Dave, I. de Pater, R. Porrata, D.-W. Kim, S.-K. King, T. Lee, H.-C. Lin, J. J. Lissauer, S. L. Marshall, P. Protopapas, J. A. Rice, M. E. Schwamb, S.-Y. Wang, C.-Y. Wen First Results From The Taiwanese-American Occultation Survey (TAOS) The Astrophysical Journal, Volume 685, Issue 2, pp. L157-L160. doi:10.1086/592741

28.2 External links

- Official website

- TAOS at National Central University, Institute of Astronomy

{{int:Coll-attribution-page|

- **Kuiper belt** *Source:* https://en.wikipedia.org/wiki/Kuiper_belt?oldid=685949021 *Contributors:* Brion VIBBER, Vicki Rosenzweig, Mav, Bryan Derksen, Robert Merkel, Berek, AstroNomer~enwiki, Jeronimo, Wayne Hardman, Andre Engels, Eob, Josh Grosse, Danny, Shsilver, XJaM, Heron, Olivier, Rickyrab, AdSR, TeunSpaans, Michael Hardy, Dhum Dhum, Minesweeper, Alfio, Egil, Card~enwiki, Looxix~enwiki, Ellywa, Cyp, Stevenj, Glenn, Andres, Jeandré du Toit, Evercat, Pizza Puzzle, Schneelocke, Hike395, The Tom, Ec5618, Timwi, Ed Cormany, Andrewman327, Nickshanks, Carbuncle, Denelson83, MK~enwiki, Robbot, Fredrik, Tomchiukc, Altenmann, Yosri, Rursus, Joelwest, Rebrane, Wikibot, Jor, Cyberia23, Cyrius, Pablo-flores, Stirling Newberry, Nephelin~enwiki, Giftlite, DocWatson42, Christopher Parham, Jyril, Harp, Geeoharee, P.T. Aufrette, Lestatdelc, Curps, Michael Devore, Avsa, Cam, Mooquackwooftweetmeow, Bolo1729, Ctachme, DNewhall, Mzajac, ScottyBoy900Q, Urhixidur, Kevyn, Ericg, Rfl, RossPatterson, Rich Farmbrough, Guanabot, Ponder, AlanBarrett, Bender235, Neil-Tarrant, Kbh3rd, Brian0918, Ylee, Kwamikagami, Downhighest, Tom, Sietse Snel, Art LaPella, RoyBoy, Iridia, Bobo192, Iamunknown, Jeffmedkeff, Tronno, Brim, Cwolfsheep, KBi, Pharos, HasharBot~enwiki, Jumbuck, Sich Bojan, Alansohn, Eric Kvaalen, Keflavich, Evil Monkey, Inge-Lyubov, Grtamlinb, Deathphoenix, Pauli133, Sturmde, Brookie, Firsfron, Uncle G, Scjessey, MONGO, Eleassar777, Sffcorgi, Steinbach, Smartech~enwiki, Palica, RedBLACKandBURN, Graham87, Mendaliv, Drbogdan, Rjwilmsi, Саша Стефановић, BlueMoonlet, Josiah Rowe, Mike s, XLerate, Bubba73, Brighterorange, Tedzsee, Vuong Ngan Ha, FlaBot, Fivemack, SchuminWeb, RobertG, RexNL, Arctic.gnome, Joedeshon, Malhonen, Chobot, Amaurea, YurikBot, Borgx, Hairy Dude, Garglebutt, Madkayaker, Fluorhydric, Shawn81, C777, Zhatt, Lusanaherandraton, Wiki alf, Lowe4091, Exir Kamalabadi, Howcheng, Semperf, Rhodekyll, Gadget850, TimeCruiserMike, Caerwine, Dan Austin, Dna-webmaster, FF2010, Ishel99, Poppy, Chesnok, TheMadBaron, Closedmouth, Nemu, Cobblet, Acer, CWenger, Nixer, Ilmari Karonen, Jasongetsdown, Co149, Serendipodous, Mikegrant, The Yeti, Luk, Deuar, Sardanaphalus, SmackBot, WilyD, Eskimbot, Ozone77, Monz, Richard B, Gilliam, Aaron of Mpls, Saros136, MK8, Foosher, DHN-bot~enwiki, Scwlong, Audriusa, Modest Genius, John Hyams, Tamfang, Mrwuggs, Vanis314, Rasimpson, T-borg, Jan.Kamenicek, J P, TTE, Ohconfucius, SashatoBot, Philrosenberg, Harryboyles, John, Jon.prasad, Writtenonsand, J 1982, Ocee, Sir Nicholas de Mimsy-Porpington, JamesFox, JorisvS, LonelyPker, Thegreatdr, Omnedon, Eurocommuter, Waggers, Dcflyer, RMHED, Daviddaniel37, Iridescent, JMK, StephenBuxton, Zero sharp, Acom, Harold f, Fenneth, Conorobradaigh, Myrrhlin, Tuvas, TheMightyOrb, Runningonbrains, Ruslik0, CuriousEric, AshLin, Outriggr (2006-2009), Ufviper, Cydebot, Lightblade, Kanags, Shirulashem, DumbBOT, Kozuch, Richhoncho, Thijs!bot, Epbr123, SchutteGod, Mbell, Headbomb, Peashy, Mentifisto, AntiVandalBot, KP Botany, Mdotley, JAnDbot, Deflective, Something14, Christopher Cooper, Rothorpe, WolfmanSF, Murgh, Bongwarrior, VoABot II, SHCarter, بلاس, Ling.Nut, Martin56, BatteryIncluded, Odros, Cpl Syx, Spellmaster, Khalid Mahmood, A2-computist, Kheider, RP88, Anaxial, Jay Litman, CommonsDelinker, Boston, J.delanoy, DrKay, AstroHurricane001, Maurice Carbonaro, Padalkar.kshitij, LordAnubisBOT, Tarotcards, ElectricValkyrie, Cometstyles, Treisijs, Jarry1250, Sage of Ice, Xiahou, Idioma-bot, Lwalt, Rathanor, VolkovBot, AlnoktaBOT, Barneca, Philip Trueman, TXiKiBoT, Jogar2, BertSen, Rei-bot, HarryAlffa, Sintaku, JhsBot, TomXP411, Rhort, Latulla, Aajacksoniv, Vimalkalyan, @pple, Upquark, Logan, Vahagn Petrosyan, NHRHS2010, EmxBot, SieBot, BotMultichill, Triwbe, RadicalOne, Murlough23, OKBot, LonelyMarble, Dabomb87, Escape Orbit, ClueBot, The Thing That Should Not Be, Eric Wester, AstroMark, Gaia Octavia Agrippa, Wysprgr2005, Mild Bill Hiccup, Drvancampen, SuperHamster, Av0id3r, Piledhigheranddeeper, Kitsunegami, Ngebendi, Njardarlogar, Scog, Ottawa4ever, JasonAQuest, BOTarate, Stepheng3, Another Believer, Gnickett1, Aitias, Cookiehead, HumphreyW, JDT1991, InternetMeme, XLinkBot, Tegles, SilvonenBot, Jk2693, Addbot, Roentgenium111, Some jerk on the Internet, DOI bot, The Other Saluton, Ronhjones, LaaknorBot, Nemo1986, LinkFA-Bot, Sardur, Tide rolls, RoosMargot, Bfigura's puppy, Lightbot, John Belushi, MuZemike, Legobot, Luckas-bot, Yobot, OrgasGirl, Guessing Game, KamikazeBot, Tbayboy, AnomieBOT, JackieBot, Aditya, ₀x, Shadowmorph, Materialscientist, Citation bot, Astor14, DynamoDegsy, ArthurBot, MauritsBot, Xqbot, Zad68, Gap9551, Ruy Pugliesi, Ataleh, RibotBOT, Triplepickle815, Ignoranteconomist, Bigger digger, Shadowjams, Elockard, Roudy66, Originalwana, Craig Pemberton, John85, Citation bot 1, Citation bot 4, Moritheil, Jonesey95, Tom.Reding, Pmokeefe, ArgGeo, Fartherred, TobeBot, GabrielEvans, Jann, Vrenator, Earthandmoon, Sideways713, RjwilmsiBot, DASHBot, EmausBot, John of Reading, Jolielegal, AbbaIkea2010, Racerx11, GoingBatty, RA0808, Jmencisom, Wikipelli, K6ka, P. S. F. Freitas, Werieth, DavidMCEddy, A2soup, Wieralee, Nascarfan1,000,000,000, Brandmeister, Donner60, Desibites, ChuispastonBot, Whoop whoop pull up, Scruffywiki, Princejack12, Fjörgynn, ClueBot NG, ATX-NL, CocuBot, MelbourneStar, This lousy T-shirt, Piast93, Jhenderson8, Movses-bot, PoqVaUSA, Rezabot, Vegeta624, Helpful Pixie Bot, Kiviuq~enwiki, Gob Lofa, Bibcode Bot, Ande0729, Jonshill, Lowercase sigmabot, CityOfSilver, George Ponderevo, Yowanvista, Altaïr, SodaAnt, Tycho Magnetic Anomaly-1, Hamish59, Cliff12345, Imgayftwyes, Yogirox234, BattyBot, Afroman123456789, Eippiepie33, Evilswann, Dexbot, Skirahm, Poop face3432, Jaredisred, Jbaugher13, Corinne, Mrjulesd, RandomLittleHelper, Reatlas, Rfassbind, Apidium23, Renerpho, Beautyon, Exoplanetaryscience, MrScorch6200, Gts-tg, Monkbot, Filedelinkerbot, Nuwanpushpakumara, Tomstar214, Tetra quark, Nichodon, LL221W, KasparBot, Proloy-Adhikary, JakeR2002, Fishyishere and Anonymous: 352

- **List of the brightest Kuiper belt objects** *Source:* https://en.wikipedia.org/wiki/List_of_the_brightest_Kuiper_belt_objects?oldid=684478700 *Contributors:* The Tom, Rursus, Urhixidur, Kwamikagami, Wavelength, Serendipodous, SmackBot, Colonies Chris, Ruslik0, Headbomb, Rothorpe, Kheider, Qwfp, Addbot, Roentgenium111, LaaknorBot, Tassedethe, 84user, Licandro, Jonesey95, Ilvon, LawBot, EmausBot, ZéroBot, H3llBot, Mjbmrbot, DoctorKubla, Rfassbind, Nichodon and Anonymous: 3

- **(19308) 1996 TO66** *Source:* https://en.wikipedia.org/wiki/(19308)_1996_TO66?oldid=681160934 *Contributors:* The Tom, JohnCastle, Jyril, Urhixidur, Discospinster, Rich Farmbrough, Kwamikagami, Iridia, RussBlau, CyberSkull, Palica, Rjwilmsi, FlaBot, TheDJ, RussBot, Martinwilke1980, Reyk, Poulpy, Serendipodous, Deuar, Sardanaphalus, Kilo-Lima, Eskimbot, Kokorik, JorisvS, Acom, Ruslik0, Headbomb, Rothorpe, Sethhater123, Kheider, Remember the dot, Anticipation of a New Lover's Arrival, The, Addbot, Captain-tucker, Lightbot, Ptbotgourou, Cflm001, Guy1890, Ipatrol, Citation bot, Astor14, Xqbot, Citation bot 1, Tom.Reding, IcesAreCool, Double sharp, DASHBot, EmausBot, GoingBatty, Solomonfromfinland, ZéroBot, ChuispastonBot, Lanthanum-138, Bibcode Bot, Rfassbind, Exoplanetaryscience, OccultZone and Anonymous: 8

- **(35671) 1998 SN165** *Source:* https://en.wikipedia.org/wiki/(35671)_1998_SN165?oldid=685991682 *Contributors:* The Tom, Rich Farmbrough, Kwamikagami, Rjwilmsi, TheDJ, Deuar, Sardanaphalus, JorisvS, Novangelis, JamesLucas, Headbomb, Rothorpe, Sethhater123, Kheider, IdLoveOne, Remember the dot, Addbot, Lightbot, Luckas-bot, Yobot, Ptbotgourou, Guy1890, Tbayboy, Tom.Reding, Ilvon, Alph Bot, DASHBot, WikitanvirBot, Hevron1998, BattyBot, Rfassbind, Exoplanetaryscience, Stamptrader and Anonymous: 1

- **38083 Rhadamanthus** *Source:* https://en.wikipedia.org/wiki/38083_Rhadamanthus?oldid=686606728 *Contributors:* Bryan Derksen, The Tom, JohnCastle, Merovingian, The Singing Badger, Urhixidur, Rich Farmbrough, Kwamikagami, Iridia, A2Kafir, Palica, Chupon, Rjwilmsi,

Mike s, FlaBot, RussBot, Ospalh, BOT-Superzerocool, Sardanaphalus, KnightRider~enwiki, Mairibot, Chris the speller, WinstonSmith, Vina-iwbot~enwiki, Ohconfucius, JorisvS, Acom, Deflective, Rothorpe, T@nn, Sethhater123, Kheider, VolkovBot, Sanddune777, DumZiBoT, Addbot, Lightbot, Luckas-bot, Xqbot, MastiBot, DASHBot, WikitanvirBot, ZéroBot, ChuispastonBot, Rfassbind, Exoplanetaryscience, DN-boards1 and Anonymous: 4

- **(15836) 1995 DA2** *Source:* https://en.wikipedia.org/wiki/(15836)_1995_DA2?oldid=673377279 *Contributors:* The Tom, Rich Farmbrough, TheDJ, Deuar, Sardanaphalus, Kokorik, JamesLucas, Sethhater123, Kheider, Remember the dot, Addbot, Lightbot, Luckas-bot, Tom.Reding, DASHBot, ChuispastonBot and OccultZone

- **(15883) 1997 CR29** *Source:* https://en.wikipedia.org/wiki/(15883)_1997_CR29?oldid=673377306 *Contributors:* The Tom, Rich Farmbrough, Kwamikagami, TheDJ, Deuar, Sardanaphalus, JamesLucas, Sethhater123, Remember the dot, MystBot, Addbot, Lightbot, LucienBOT, ZéroBot and Fjörgynn

- **(79978) 1999 CC158** *Source:* https://en.wikipedia.org/wiki/(79978)_1999_CC158?oldid=673443968 *Contributors:* The Tom, Rich Farm-brough, Kwamikagami, TheDJ, Deuar, Sardanaphalus, JorisvS, Ruslik0, JamesLucas, Sethhater123, Kheider, Remember the dot, Addbot, Lightbot, Yobot, Tom.Reding, DASHBot, WikitanvirBot, Carrotou, Stamptrader, Tommaso Vizzaccaro and Anonymous: 1

- **(182294) 2001 KU76** *Source:* https://en.wikipedia.org/wiki/(182294)_2001_KU76?oldid=662208391 *Contributors:* The Tom, Phil Boswell, Rjwilmsi, Cesium 133, JorisvS, Kheider, Remember the dot, Addbot, Tom.Reding, Tim1357, DASHBot, ZéroBot and OccultZone

- **2003 LA7** *Source:* https://en.wikipedia.org/wiki/2003_LA7?oldid=678624179 *Contributors:* Rich Farmbrough, JorisvS, Ruslik0, Kheider, Remember the dot, Lightbot, Ptbotgourou, WebCiteBOT, Io Herodotus, Tom.Reding, Double sharp, DASHBot, Solomonfromfinland, KLBot2, Exoplanetaryscience and Anonymous: 1

- **(136120) 2003 LG7** *Source:* https://en.wikipedia.org/wiki/(136120)_2003_LG7?oldid=665417311 *Contributors:* Danny, The Tom, Merovin-gian, Rich Farmbrough, Koavf, MapsMan, Sardanaphalus, SmackBot, JorisvS, CmdrObot, Ruslik0, Sethhater123, Kheider, Remember the dot, DumZiBoT, Addbot, Lightbot, JackieBot, Tom.Reding, Double sharp and Solomonfromfinland

- **2010 FX86** *Source:* https://en.wikipedia.org/wiki/2010_FX86?oldid=678103678 *Contributors:* Bearcat, JorisvS, Kheider, Katharineamy, Ad-dbot, Guy1890, Tom.Reding, RedBot, Solomonfromfinland, Pluto and Beyond, Rfassbind, Exoplanetaryscience, GENVELES, DN-boards1 and Anonymous: 1

- **2010 RF43** *Source:* https://en.wikipedia.org/wiki/2010_RF43?oldid=680459652 *Contributors:* Bearcat, JorisvS, Katharineamy, Guy1890, Tom.Reding, Solomonfromfinland, Hevron1998, Gob Lofa, BattyBot, Pluto and Beyond, Rfassbind, Machdelu and Exoplanetaryscience

- **2013 FZ27** *Source:* https://en.wikipedia.org/wiki/2013_FZ27?oldid=688179044 *Contributors:* Kwamikagami, JorisvS, Kheider, Ambi Valent, Roentgenium111, Guy1890, Tom.Reding, Exoplanetaryscience, GENVELES, Deneb in Cygnus and Anonymous: 2

- **2014 MT69** *Source:* https://en.wikipedia.org/wiki/2014_MT69?oldid=683528221 *Contributors:* Kolbasz, JorisvS, Yobot, Tom.Reding, Obankston, Renerpho, LL221W and Deneb in Cygnus

- **2014 MU69** *Source:* https://en.wikipedia.org/wiki/2014_MU69?oldid=689362082 *Contributors:* Pmsyyz, Zaslav, Kwamikagami, Jonathunder, Drbogdan, Bubba73, Kolbasz, Reyk, Ericl, JorisvS, DmitTrix, Dawkeye, BatteryIncluded, Kheider, Njardarlogar, Astrofreak92, AnomieBOT, Arsia Mons, Fotaun, Io Herodotus, Tom.Reding, IJBall, Obankston, Solomonfromfinland, Wieralee, Matsaball, BG19bot, Rfassbind, Renerpho, Exoplanetaryscience, Monkbot, Bodhisattwa, LL221W, Leo221W, Deneb in Cygnus and Anonymous: 8

- **2014 OS393** *Source:* https://en.wikipedia.org/wiki/2014_OS393?oldid=683527977 *Contributors:* Kolbasz, Tom.Reding, Obankston, Rener-pho, LL221W and Deneb in Cygnus

- **2014 PN70** *Source:* https://en.wikipedia.org/wiki/2014_PN70?oldid=683716644 *Contributors:* Kolbasz, Ericl, JorisvS, Arsia Mons, Tom.Reding, Obankston, Renerpho, LL221W, Deneb in Cygnus and Anonymous: 1

- **2014 UM33** *Source:* https://en.wikipedia.org/wiki/2014_UM33?oldid=686606534 *Contributors:* Rpyle731, Yobot, Io Herodotus, Wgolf, Rfassbind, Exoplanetaryscience, DN-boards1 and Deneb in Cygnus

- **2011 KW48** *Source:* https://en.wikipedia.org/wiki/2011_KW48?oldid=678071098 *Contributors:* JorisvS, AnomieBOT, Io Herodotus, Tom.Reding, IJBall, Obankston, Winner 42, Solomonfromfinland, Parcly Taxel, Exoplanetaryscience, Verdana Bold, Deneb in Cygnus and Anonymous: 4

- **(230965) 2004 XA192** *Source:* https://en.wikipedia.org/wiki/(230965)_2004_XA192?oldid=680457057 *Contributors:* Kwamikagami, Gre-gorB, RussBot, Omega13a, JorisvS, Dl2000, Mild Bill Hiccup, Addbot, Lightbot, Tom.Reding, Double sharp, DASHBot, Stormchaser89, ZéroBot, Hevron1998, ChuispastonBot, Gob Lofa, Rfassbind and Anonymous: 1

- **(420356) 2012 BX85** *Source:* https://en.wikipedia.org/wiki/(420356)_2012_BX85?oldid=685807770 *Contributors:* Kwamikagami, Wiki-mandia, Ilvon, Double sharp, Rfassbind, Exoplanetaryscience, DN-boards1 and Deneb in Cygnus

- **Actaea (moon)** *Source:* https://en.wikipedia.org/wiki/Actaea_(moon)?oldid=689413512 *Contributors:* The Tom, Kwamikagami, Pegship, WolfmanSF, AnomieBOT, Robvanvee, BG19bot, Rfassbind, Gulaschkanone99 and Anonymous: 1

- **Heliosphere** *Source:* https://en.wikipedia.org/wiki/Heliosphere?oldid=687453167 *Contributors:* Bryan Derksen, Darkwind, Julesd, The Tom, SEWilco, Kjell André, Jredmond, Alba, David Gerard, Giftlite, Awolf002, Hargettp, Alison, Rookkey, Beland, Discospinster, Vsmith, Ben-der235, RJHall, Svdmolen, Gedanken, Evolauxia, Dreish, Cohesion, Alansohn, Ungtss, SemperBlotto, Splat, Mysdaao, Wsloand, Kitch, DrDave-HPP, BillC, MONGO, Tabletop, Eilthireach, SDC, BD2412, JIP, Rjwilmsi, Nightscream, Marasama, Bubba73, Nandesuka, SchuminWeb, Who, Leslie Mateus, Kolbasz, Gareth E Kegg, Chobot, Tene, Bgwhite, Banaticus, YurikBot, Borgx, Tehnakki, Gaius Cornelius, Bovineone, SFC9394, Evrik, Phenz, Super jedi droid, Thnidu, Exodio, Serendipodous, Finell, AndrewWTaylor, Sardanaphalus, SmackBot, Mscuthbert, Valley2city, Bluebot, Colonies Chris, BIL, Acdx, Ligulembot, TiCPU, Michael Slana, JorisvS, Danilot, Kencf0618, Joseph Solis in Australia, FairuseBot, Daggerstab, Leevanjackson, Drinibot, Ruslik0, Phatom87, User6985, Themightyquill, Cydebot, Empln, O, Mikeeg555, 49oxen, SadanYagci, Flex Flint, JAnDbot, Owenozier, Gavia immer, Roidroid, Brett, WolfmanSF, Swpb, BatteryIncluded, Cpl Syx, Mynameisntbob1, Kheider, Keithmanders, Axemanstan, R'n'B, Hans Dunkelberg, Maurice Carbonaro, Acalamari, Xbspiro, Lunokhod, Jdemp42, Idioma-bot, Bobpowell1, Mathwhiz 29, Sean D Martin, Broadbot, Fbs. 13, BotKung, Aalox, SieBot, Sakkura, Aureliusweb, Jdaloner, Nipsonanomhmata,

Explicit, Martarius, ClueBot, EoGuy, Easphi, MicroVirus, Solar-Wind, Awickert, Soxrule21, Eitheta, Sallicio, Jonverve, Doglet1, Venyx, Addbot, Basilicofresco, Roentgenium111, DOI bot, Cuaxdon, Download, Tassedethe, Luckas-bot, Yobot, Bunnyhop11, Lairdwh, AnomieBOT, Wtachi, JackieBot, Piano non troppo, Hunnjazal, Citation bot, ArthurBot, DirlBot, Marshallsumter, Xqbot, Quazgaa, Topherwhelan, Kompar~enwiki, Fotaun, FrescoBot, Deltik, Paine Ellsworth, Originalwana, EmilTyf, Citation bot 1, Pinethicket, Tom.Reding, Full-date unlinking bot, Stardust1043, Canuck100, James Gunasekera, RjwilmsiBot, EmausBot, WikitanvirBot, Helium4, Frendinius, Tommy2010, Mmeijeri, A2soup, Dkevanko, Ego White Tray, Tot12, ChuispastonBot, Mappetop, ClueBot NG, AlbertBickford, Sleddog116, Frietjes, Mmarre, Gob Lofa, Bibcode Bot, BG19bot, Avinash.royyuru, Zedshort, Filiosus's Saga, BattyBot, Colinphilipjohnstone, Justincheng12345-bot, Qxukhgiels, Dexbot, Hmainsbot1, Mogism, Yanpas, Doggy23woggy4, Reatlas, Rfassbind, Ruby Murray, Wacgyver, Fginc, Alanwpeck, Brown43, Pongox, Elenceq, Monkbot, Filedelinkerbot, DSCrowned, Tetra quark, Isambard Kingdom, KasparBot, Kiwifist and Anonymous: 145

- **Oort cloud** *Source:* https://en.wikipedia.org/wiki/Oort_cloud?oldid=689271668 *Contributors:* AxelBoldt, Paul Drye, Brion VIBBER, Mav, Bryan Derksen, The Anome, Manning Bartlett, -- April, Eob, XJaM, Chrislintott, Jsc1973, Heron, Arj, Nealmcb, Patrick, Ixfd64, Paul A, Alfio, Looxix~enwiki, Muriel Gottrop~enwiki, Glenn, Sugarfish, Andres, Jeandré du Toit, Evercat, Schneelocke, Hike395, Barak~enwiki, The Tom, Lfh, Jwrosenzweig, Tpbradbury, Paul-L~enwiki, Carbuncle, Robbot, Sander123, Goethean, Yosri, Rursus, Anthony, Jor, Cyrius, Dina, Craig Butz, Nephelin~enwiki, Pabouk, Giftlite, Awolf002, Jyril, Eran, Bradeos Graphon, Everyking, No Guru, Curps, Utcursch, Antandrus, HorsePunchKid, Satori, ScottyBoy900Q, CesarFelipe, NoPetrol, Neutrality, Urhixidur, Joyous!, Eisnel, Mike Rosoft, Discospinster, Rich Farmbrough, Guanabot, Vsmith, Pie4all88, Ponder, Bender235, Rubicon, ESkog, Cyclopia, Dpotter, Tompw, RJHall, Mr. Billion, Kwamikagami, RoyBoy, RTucker, Bobo192, Pyramide~enwiki, La goutte de pluie, Ardric47, Jonathunder, Nsaa, Eddideigel, HasharBot~enwiki, Orangemarlin, Alansohn, Axl, Water Bottle, M3tainfo, Evil Monkey, Bsadowski1, Pauli133, SCEhardt, Yury Tarasievich, SDC, Smartech~enwiki, Graham87, Marskell, Qwertyus, Drbogdan, Rjwilmsi, Trlovejoy, Mike s, Brighterorange, Watcharakorn, FlaBot, RobertG, Arctic.gnome, Chobot, Sharkface217, DVdm, Peter Grey, YurikBot, Wester, Hairy Dude, Jimp, Phantomsteve, Fluorhydric, NawlinWiki, Dysmorodrepanis~enwiki, Wiki alf, Justin Eiler, Kdbuffalo, Tony1, Dna-webmaster, Miraculouschaos, Sandstein, StuRat, Mike Dillon, SFGiants, Jmackaerospace, Wechselstrom, Kier07, Tzepish, NeilN, Jasongetsdown, Serendipodous, Sardanaphalus, SmackBot, Melchoir, Cthompson, Wegesrand, WilyD, Brossow, Kintetsubuffalo, Yamaguchi⚀⚀, Onsly, Peter Isotalo, Gilliam, Skizzik, Saros136, Qasrani, MK8, MalafayaBot, Hibernian, Sadads, Modest Genius, Robogun, Tamfang, Benjamin Mako Hill, GrahameS, Wen D House, Infovoria, DMacks, Samuel Sol, SashatoBot, TKarrde, Vgy7ujm, J 1982, Ocee, JorisvS, Mgiganteus1, IronGargoyle, Ckatz, A. Parrot, MarkSutton, Dr Smith, SandyGeorgia, Saxbryn, Marshall Stax, Iridescent, Michaelbusch, Paul venter, Newone, Civil Engineer III, Harold f, Xcentaur, AarrowOM, Blue Mirage, CmdrObot, Tanthalas39, Vyznev Xnebara, ThreeBlindMice, Ruslik0, CuriousEric, Burkedavis, Standonbible, Safalra, Funnyfarmofdoom, Icarus of old, Gmcomp, Cydebot, Gogo Dodo, DumbBOT, Ian Macintosh, JodyB, Gimmetrow, Epbr123, Wikid77, Qwyrxian, Markus Pössel, Headbomb, Tonyle, Marek69, Davidhorman, The Proffesor, CielProfond, Escarbot, QuiteUnusual, Quintote, JAnDbot, Deflective, MER-C, John a s, Rothorpe, Acroterion, Penubag, Magioladitis, WolfmanSF, Murgh, JNW, Midgrid, Gabriel Kielland, BatteryIncluded, DerHexer, Megalodon99, Kheider, An Sealgair, MartinBot, RP88, Gsrdzl, CommonsDelinker, Vox Rationis, Tgeairn, J.delanoy, DrKay, AstroHurricane001, Ayecee, TomS TDotO, Darth Mike, Medium69, Katalaveno, Tarotcards, Dishmaster, Gurchzilla, ElectricValkyrie, AntiSpamBot, Amteixeira, Kw0134, STBotD, Josh Tumath, Azmodes, Jparenti, DorganBot, Treisijs, Xiahou, Idioma-bot, Ottershrew, VolkovBot, Midoriko, RingtailedFox, Jeff G., AlnoktaBOT, Wild Deuce, TXiKiBoT, GimmeBot, Rei-bot, Anonymous Dissident, How M, JhsBot, UnitedStatesian, Cosmicos, Spiral5800, SheffieldSteel, James McBride, Burntsauce, Rootmoose, Piperdown, Ceranthor, AlleborgoBot, NHRHS2010, Hrafn, SieBot, Sonicology, N-HH, Smsarmad, Lambermo, Antonio Lopez, Goustien, Lightmouse, Murlough23, Ealdgyth, Kwib, Mcpx, BenoniBot~enwiki, OKBot, LonelyMarble, Randomblue, Dabomb87, Twinsday, Elassint, ClueBot, LAX, Timeineurope, AstroMark, LizardJr8, Piledhigheranddeeper, Solar-Wind, Kitsunegami, Excirial, Jusdafax, Cenarium, Njardarlogar, 7&6=thirteen, Chaosdruid, Michael6633, Aitias, Nabs2597, Jeffrey Wordsmith, TimothyRias, InternetMeme, Bearsona, Pichpich, Ost316, WikiDao, ElMeBot, Addbot, DOI bot, Ronhjones, G.Hagedorn, Leszek Jańczuk, Download, Protonk, Glane23, LinkFA-Bot, Setwisohi, Ace45954, Tide rolls, Lightbot, Бiлецький B.C., Winston, Legobot, Yinweichen, Luckas-bot, Yobot, ET202, AnakngAraw, Azcolvin429, AnomieBOT, Floquenbeam, Materialscientist, Citation bot, DirlBot, LilHelpa, Xqbot, Sionus, Braxtonw1, Gap9551, Metafax1, GrouchoBot, Omnipaedista, RibotBOT, Kyng, Ymejer, Rb88guy, Bigger digger, Erik9, Fotaun, ⚀⚀, Surv1v4l1st, Paine Ellsworth, Io Herodotus, Sémaphore, Citation bot 1, Kobrabones, Cubs197, Pinethicket, Degen Earthfast, IMTHIYAS ABDULLA, Tom.Reding, Serols, Logical Gentleman, IVAN3MAN, TobeBot, Nilock, Vrenator, MikeLousado, Zachareth, Tbhotch, Sideways713, RjwilmsiBot, EmausBot, CTtcg, AbbaIkea2010, Racerx11, Britannic124, Solomonfromfinland, MithrandirAgain, Kiwi128, Quondum, EricWesBrown, L1A1 FAL, Old-timer0, S Whistler, Donner60, Raszoo, ChuispastonBot, AndyTheGrump, I hate whitespace, Llightex, ClueBot NG, Kiruria281, Freiza667, Karuba333, Helpful Pixie Bot, Rsercher, Titodutta, Gob Lofa, Bibcode Bot, Lowercase sigmabot, BG19bot, Ninney, Rtinetti, RememberThem, Cliff12345, Klilidiplomus, RichardLyre, CarnivorousGnomeCatuse, Tonyxc600, The Illusive Man, ChrisGualtieri, Dexbot, Pluto and Beyond, Leptus Froggi, Tony Mach, RotlinkBot, Rfassbind, PhoBo, Cumbril, Exoplanetaryscience, RhinoMind, BlueAzulon, Mahusha, Zachariah1978, Monkbot, CraigyDavi, Killieo999, Tetra quark, Deneb in Cygnus, Elisa.arizono and Anonymous: 357

- **Oort limit** *Source:* https://en.wikipedia.org/wiki/Oort_limit?oldid=668411420 *Contributors:* Jengod, The Tom, Alvestrand, RJHall, VoidLurker, Solar-Wind, MystBot, Addbot, Midas02 and Exoplanetaryscience

- **Scattered disc** *Source:* https://en.wikipedia.org/wiki/Scattered_disc?oldid=679698087 *Contributors:* Bryan Derksen, Egil, Jimfbleak, The Tom, Wikiborg, Colipon, Timc, IceKarma, CalRis25, Jeffq, Phil Boswell, PedroPVZ, Xanzzibar, Filemon, Ancheta Wis, Awolf002, Jyril, Wolfkeeper, Xerxes314, Curps, Rich Farmbrough, Florian Blaschke, D-Notice, Bender235, RJHall, Kwamikagami, Art LaPella, The shaggy one, Gunter.krebs, Jared81, Cromwellt, Pauli133, Pytom, Benhocking, SDC, Smartech~enwiki, Whoutz, Rjwilmsi, Oblivious, FlaBot, RobertG, Nihiltres, Chobot, YurikBot, Wavelength, Borgx, Jimp, Wimt, Zwobot, Merosonox, Silverhill, Dna-webmaster, Poppy, RMeier, Reyk, Ilmari Karonen, Abhishekmathur, Serendipodous, Deuar, Sardanaphalus, SmackBot, Herostratus, Eskimbot, John Hyams, WinstonSmith, BIL, Jan.Kamenicek, Ultraexactzz, J P, Samuel Sol, Ohconfucius, JorisvS, Hans van Deukeren, Erwin, Eurocommuter, SandyGeorgia, Jggouvea, Peyre, Laddiebuck, Joseph Solis in Australia, Carroy~enwiki, Dia^, Ruslik0, Heatsketch, Ufviper, Dominicanpapi82, Richhoncho, Mattisse, Thijs!bot, Headbomb, Mrshaba, JAnDbot, Something14, Igodard, East718, Rothorpe, JKaspar, Penubag, WolfmanSF, Murgh, VoABot II, Ling.Nut, Perebot~enwiki, GregU, Kheider, DGG, Archolman, CommonsDelinker, DrKay, Tarotcards, Chiswick Chap, Robertgreer, Lamp90, Treisijs, CardinalDan, VolkovBot, RingtailedFox, Quentonamos, TXiKiBoT, GimmeBot, A4bot, Anna Lincoln, Bob Andolusorn, BotKung, Geometry guy, Nibios, Ceranthor, Sources said, Jayhawks Lover, BotMultichill, Jbmurray, Ealdgyth, LonelyMarble, Anchor Link Bot, Randomblue, Nergaal, ClueBot, Artichoker, The Thing That Should Not Be, Frmorrison, Fenwayguy, Piledhigheranddeeper, Anon lynx, PixelBot, Njardarlogar, Calor, Dana boomer, RexxS, Editorofthewiki, ZooFari, Addbot, Willking1979, DOI bot, 84user, Ace45954, Zorrobot, Ettrig,

Luckas-bot, Yobot, AnomieBOT, Piano non troppo, Citation bot, Natural RX, Xqbot, Gap9551, FrescoBot, Dogposter, Citation bot 1, Ko-brabones, Jonesey95, Tom.Reding, Gruntler, Ilvon, RandomStringOfCharacters, C messier, FoxBot, ATBS, Dinamik-bot, GGT, AMN3SiA, Earthandmoon, Azninvasion344, Turkeyleg778, Hans Langschwanz ist wieder da, Computergeek9911, Goodboybad123, Sideways713, TjBot, Ripchip Bot, DASHBot, EmausBot, ZéroBot, H3llBot, Fjörgynn, ClueBot NG, Rezabot, Helpful Pixie Bot, Gob Lofa, Bibcode Bot, Zig-gie456789, Dexbot, Br'er Rabbit, Tony Mach, Rfassbind, Whyasknow, Kogge, Monkbot and Anonymous: 64

- **Taiwanese–American Occultation Survey** *Source:* https://en.wikipedia.org/wiki/Taiwanese%E2%80%93American_Occultation_Survey?oldid=675165807 *Contributors:* The Tom, Rjwilmsi, Serendipodous, Egsan Bacon, Dicklyon, Kheider, CommonsDelinker, KylieTastic, Cosmo0, Fedhere, D.i.l., Mortense, Tom.Reding, Skyerise, EmausBot, Lucas Thoms, Fjörgynn, Helpful Pixie Bot, Guadalupehat, Kcw19, Tianxuan777 and Anonymous: 4

- **File:(19308)_1996_TO66_imaged_by_NTT_cut_out.jpg** *Source:* https://upload.wikimedia.org/wikipedia/commons/5/53/%2819308%29_1996_TO66_imaged_by_NTT_cut_out.jpg *License:* CC BY 4.0 *Contributors:*

- image *Original artist:* ESO

- **File:14-281-KuiperBeltObject-ArtistsConcept-20141015.jpg** *Source:* https://upload.wikimedia.org/wikipedia/commons/9/98/14-281-KuiperBeltObject jpg *License:* Public domain *Contributors:* http://www.nasa.gov/sites/default/files/14-281_0.jpg *Original artist:* ASA, ESA, and G. Bacon (STScI)

- **File:2003EL61art.jpg** *Source:* https://upload.wikimedia.org/wikipedia/commons/9/90/2003EL61art.jpg *License:* Public domain *Contributors:* [1] *Original artist:* A. Feild (Space Telescope Science Institute)

- **File:2003_UB313_near-infrared_spectrum.png** *Source:* https://upload.wikimedia.org/wikipedia/commons/d/d6/2003_UB313_near-infrared_spectrum.png *License:* CC-BY-SA-3.0 *Contributors:* ? *Original artist:* ?

- **File:2014_MU69_Discovery_Images_Animated.gif** *Source:* https://upload.wikimedia.org/wikipedia/commons/f/f9/2014_MU69_Discovery_Images_Animated.gif *License:* Public domain *Contributors:*

- http://www.planetary.org/multimedia/space-images/small-bodies/pt1-discovery.html *Original artist:* NASA, ESA, SwRI, JHU/APL, and the New Horizons KBO Search Team

- **File:2014_MU69_orbit.jpg** *Source:* https://upload.wikimedia.org/wikipedia/commons/5/53/2014_MU69_orbit.jpg *License:* Public domain *Contributors:*

- http://astronomynow.com/2015/08/28/nasas-new-horizons-team-selects-potential-kuiper-belt-flyby-target/ *Original artist:* Alex Parker

- **File:2014_MU69_size_comparison_01.jpg** *Source:* https://upload.wikimedia.org/wikipedia/commons/f/f5/2014_MU69_size_comparison_01.jpg *License:* Public domain *Contributors:* http://pluto.jhuapl.edu/News-Center/PI-Perspectives/images/10-23-2014-Compare-PT1-on-Earth.jpg *Original artist:* New Horizons team

- {{int:Coll-image-attribution|File:951_Gaspra.jpg|https://upload.wikimedia.org/wikipedia/commons/8/81/951_Gaspra.jpg|Public domain|Cropped from TIFF image from [http://www.solarviews.com/cap/ast/gaspra3.htm Solarviews.com|NASA}}

- **File:Ambox_important.svg** *Source:* https://upload.wikimedia.org/wikipedia/commons/b/b4/Ambox_important.svg *License:* Public domain *Contributors:* Own work, based off of Image:Ambox scales.svg *Original artist:* Dsmurat (talk · contribs)

- **File:Comet_Hale-Bopp.jpg** *Source:* https://upload.wikimedia.org/wikipedia/commons/2/2f/Comet_Hale-Bopp.jpg *License:* CC-BY-SA-3.0 *Contributors:* From English Wikipedia (the same filename) *Original artist:* w:User:Mkfairdpm from English Wikipedia

- **File:Commons-logo.svg** *Source:* https://upload.wikimedia.org/wikipedia/en/4/4a/Commons-logo.svg *License:* ? *Contributors:* ? *Original artist:* ?

- **File:Crab_Nebula.jpg** *Source:* https://upload.wikimedia.org/wikipedia/commons/0/00/Crab_Nebula.jpg *License:* Public domain *Contributors:* HubbleSite: gallery, release. *Original artist:* NASA, ESA, J. Hester and A. Loll (Arizona State University)

- **File:Dust_Models_Paint_Alien'{}s_View_of_Solar_System.ogv** *Source:* https://upload.wikimedia.org/wikipedia/commons/8/82/Dust_Models_Paint_Alien%27s_View_of_Solar_System.ogv *License:* Public domain *Contributors:* [http://www.nasa.gov/multimedia/videogallery/index.html?media_id=18410668 NASA Multimedia] *Original artist:* NASA

- **File:Eris_and_dysnomia2.jpg** *Source:* https://upload.wikimedia.org/wikipedia/commons/5/5b/Eris_and_dysnomia2.jpg *License:* Public domain *Contributors:* http://www.nasa.gov/mission_pages/hubble/news/eris.html . Originally uploaded to en.wikipedia by en:User:Serendipodous. For more information, see the description page *Original artist:* NASA, ESA, and M. Brown

- **File:Evolving_Heliophysics_System_Observatory.jpg** *Source:* https://upload.wikimedia.org/wikipedia/commons/d/d5/Evolving_Heliophysics_System_Observatory.jpg *License:* Public domain *Contributors:* NASA Little SDO Facebook *Original artist:* NASA

- **File:ExampleUpdatedHistogramOfTNOsemimajoraxii.png** *Source:* https://upload.wikimedia.org/wikipedia/en/6/68/ExampleUpdatedHistogramOfTNC png *License:* CC-BY-SA-3.0 *Contributors:*

Data from the minor planet center, plotted in GNUplot
Previously published: I ain't done that.

Original artist:
WilyD

- **File:Folder_Hexagonal_Icon.svg** *Source:* https://upload.wikimedia.org/wikipedia/en/4/48/Folder_Hexagonal_Icon.svg *License:* Cc-by-sa-3.0 *Contributors:* ? *Original artist:* ?

- http://nssdc.gsfc.nasa.gov/database/MasterCatalog?sc=1999-003A *Original artist:* NASA

|

- Creative Commons Attribution-Share Alike 3.0

}}

www.ingramcontent.com/pod-product-compliance
Lightning Source LLC
Chambersburg PA
CBHW080820180526
45168CB00006B/2522